WHERE BRIDGES STAND

THE RIVER LEE BRIDGES OF CORK CITY

ANTÓIN O'CALLAGHAN

The History Press

First published 2012
Reprinted 2022

The History Press
97 St George's Place,
Cheltenham, Gloucestershire GL50 3QB
www.thehistorypress.co.uk

British Library Cataloguing in Publication Data.
A catalogue record for this book is available from the British Library.

ISBN 978 1 84588 746 9

Typesetting and origination by The History Press
Printed by TJ Books Limited, Padstow, Cornwall

CONTENTS

ACKNOWLEDGEMENTS

Many years ago, while living away from Cork, I was asked how many bridges span the twin channels of the Lee that enclose the island centre of the city. Unable to answer the question definitively, I asked my father, a man who had spent over fifty years walking to his workplace in the city centre from his north-side home. He too was unable to give a definite answer. Thus began a journey on which I encountered not just the story of Cork's bridges, but the history of the city itself. On that journey I met a number of people that today I am proud to call my friends – fellow travellers with a passion for the history of their native place. I want to thank Ronnie Herlihy, Tom Spalding, Dr Colin Rynne, Ciara Brett, and Diarmuid O'Drisceoil for all their help and encouragement. I also want to thank the staff of the Local Studies Department in the Cork City Library for all their assistance, especially Kieran, Mary, Caroline, Paul, Stephen, John and Eamonn. Special thanks to my friend and colleague Paul O'Flynn for all his advice and support and also to Dr John Borgonovo, who is by now fully qualified to be classed amongst those that consider the study of Cork the study of their native place. His encouragement and enthusiasm were very much appreciated.

I cannot thank Dr Dónal Ó Drisceoil enough. His knowledge of Cork's history, his interest in my work, and his advice on everything from source material to photographic records was second to none. I really appreciate his encouragement and I continue to value his friendship and advice. *Buíochas mór duit a Dhónal.* It is not an exaggeration to say that this work would not have been completed without the friendship, help and encouragement of two great Corkmen, Michael Lenihan and Pat Poland. Michael, through his great knowledge and his private collection of Cork historical memorabilia, has made a huge contribution to this work. So too has Pat, with his immense knowledge of almost every aspect of Cork history. Next one's on me lads.

I also want to acknowledge the advice and support given to me by Ronan Colgan and all of the team at The History Press Ireland. Their enthusiasm for the project was a driving force behind its completion. Finally, I want to thank my family for living with the history of Cork's bridges for almost thirty years now. To my children, Lorna and Brenton, your support, not least in historical and technical matters, was invaluable. To the love of my life, my wife Sandra, your encouragement, support and love helped me to fulfil another dream. My thanks and love always.

And so to the bridges …

INTRODUCTION

> The physical landscape with its building types, road networks and farms, combined
> with the mental landscape created by place-names and customs, become one method
> of understanding how people in the past lived out their daily lives in local places.[1]

In November 1984, Lord Mayor Liam Burke performed the opening ceremony of
the two newest bridges that had been built across the North and South Channels
of the River Lee, between which sits the island centre of the City of Cork. They were
named the Éamon de Valera and Michael Collins bridges after two heroes of the War
of Independence which led to the emergence of Free-State Ireland. Collins and de
Valera, were on opposite sides of the treaty divide that led ultimately to Civil War,
during which, in August 1922, Michael Collins was fatally shot in west Cork. Éamon
de Valera went on to become Taoiseach of the country in 1932 and subsequently
became President of the Republic of Ireland in 1959, serving two seven-year terms.

The names of Collins and de Valera, and the politics that they represent,
became associated with what was known as 'Civil War politics' in the country: the
seemingly irreconcilable difference between those who sought a thirty-two-county
Ireland and those who, while never denying this as a long-term aspiration, were
content to have the twenty-six-county nation that emerged from the treaty of 1922
as a stepping stone to complete independence. It was therefore a symbolic look to
the future when, in 1984, the two bridges, which formed part of a single project
in Cork's development at that time, were named in honour of these two statesmen.
The Michael Collins and Éamon de Valera bridges stood as symbols of the end of a
half-century of Civil-War differences: a unifying of previously diverse viewpoints
in pursuit of a better future. Another significance was that these were the last two
bridges in the city to be given names with political overtones.[2] Subsequent bridges
were given non-political names such as St Finbarr's, Shandon and Mardyke.

Elsewhere too, the naming of bridges was undergoing change. In the nation's
capital city of Dublin, nineteenth-century bridges had been named after political
figures such as the Earl of Carlisle, the Duke of Ormond and the English Monarch in
the form of Queen's Bridge. These were subsequently altered as part of the nationalist
agenda to commemorate people such as Daniel O'Connell, O'Donovan Rossa and

Drawing of Proby's Bridge and old St Finbarr's Cathedral by James Brennan.

Liam Mellows. However, the newest, twenty-first-century bridges in Dublin were named after the writers James Joyce and Samuel Beckett and the New Millenium Bridge marked the year 2000.[3] In Limerick, what had previously been Wellesley Bridge became Sarsfield Bridge, and the next crossing down-stream from this one, which opened in May 1988, was at first commonly known as the New Bridge and later was named the Shannon Bridge.

The naming conventions employed for civic infrastructural developments are an important element in the narrative of the identity of a place. They are records of historical events and contexts, not just at the time of construction but in subsequent times when changes in societal aspiration can lead these developments being renamed.

The rituals performed at the opening ceremonies of civic developments are also significant as they are a means of advancing the agendas, be they political or apolitical, of the power-brokers of the day.

The fact that bridges are infrastructural projects and developments, lead us to look at the purpose of their construction. Why were they built and what effect did they have on the development of the surrounding area?

In his study of *Famous Bridges of the World*, bridge-builder and historian David B. Steinman wrote, 'once you begin to notice bridges, really look at them and not just cross over them with your mind on something else, you see that they are not all alike, that there are different kinds of bridges'.[4]

Throughout Ireland, there are a variety of bridge types, from beam-support and arch to suspension bridges, made variously of timber, iron, and stone. Like the naming conventions employed, the architecture and design of the different bridges not only tell us of the bridge-builders themselves, but they contribute to the identity of the places in which they were built.

Although the primary function of a bridge remains, of course, the joining together of otherwise disparate geographical parts, many other inferences can be drawn from such structures.

It can be suggested therefore that there are a number of defining characteristics of a bridge. Firstly, by interconnecting two spaces – often by traversing a hazard – a bridge facilitates social and economic interactivity. In achieving this, mankind's engineering ingenuity is employed and so a bridge is a man-made response to the obstacles and challenges of the natural environment. Secondly, apart from the social and economic benefits, bridge-building results in the architectural adornment of the hinterland. Thirdly, a bridge is a structural testimony to the history of the area and thereby spanning, if you will, both space and time. Finally, a bridge can be the stage upon which the power brokers can enact their power rituals.

Writing in October 1998, David Bennett, author of *The Creation of Bridges*, made three valid and interesting points. He said that bridges, 'are the umbilical cords of humankind's progress over the centuries'. All bridges, regardless of type, 'have the same common goal of serving the need for better access, better transportation links and trade between local communities and international boundaries'. Again regardless of bridge type, 'every bridge project must start with a vision of its creation, followed by the endeavour to make that creation a reality'.[5]

Many great works have been written on the subject of bridges, Steinman's *Great Bridges of the World* and Bennett's *The Creation of the Bridges* being amongst the most notable. Informative as both books are, and spectacularly illustrated in the case of Bennett's, they deal with the great bridges of the world, such as the Golden Gate of Francisco, the Forth Rail Bridge in Edinburgh and the Severn Bridge between England and Wales. The accounts and stories of the thousands of less spectacular bridges that service communities and places throughout the world are left to the realms of engineering journals and local history books.

In this regard, Ireland is particularly well-served by one publication from 1991.[6] O'Keefe and Simmington's *Irish Stone Bridges, History and Heritage* tells the story of bridges from about AD 1000 and is a study of Ireland's old stone road bridges: their history, function and design. Another worthy addition to the list of publications about Ireland's bridges is *Across Deep Waters* by chartered civil engineer Michael Barry.[7] While these books deal with Irish bridges across the entire island, bridges on individual river systems or local areas are recounted in papers such as that of Phillips and Hamilton for the river Liffey in Dublin or the *Bridges of County Offaly* by Fred Hammond, as well as a variety of other articles and essays that can be located in the local libraries throughout the country.[8]

South Gate Bridge from the Signal Series. (Courtesy Michael Lenihan.)

There have been a number of accounts of the river bridges of Cork City published in recent years. *Of Timber, Iron and Stone* was a short account published in 1990 which sought to answer the simple question of the number of bridges that spanned the twin channels of the Lee at that time. Some information in that account was sourced from McNamara's *Portrait of Cork*, a work which dealt with all of the architectural features in Cork City and the part that bridges played in the industrial archaeology of the city is dealt with in Colin Rynne's *The Industrial Archaeology of Cork City and its Environs.*

Given the topography of Cork's island centre, most general histories make reference to the part played by bridges in the city's development. Lewis' *Topographical Dictionary of Cork* and Crofton Croker's *Researches in the South of Ireland*, both published in the early nineteenth century, have sections that deal with the city's bridges. Gibson and Windelle, in their histories, follow suit. In the modern era, historians such as John McCarthy and Sean Beecher, as well as McNamara, O'Callaghan and Rynne mentioned above, have each built upon previous works so that the basic story of Cork's bridges has been chronicled. However, while *Of Timber, Iron and Stone* is the only work that deals exclusively with the river bridges of Cork, none of the works take the defining characteristics outlined earlier in this introduction as a basis for a study of Cork's River Lee bridges. The purpose of this book is to expand upon the 1990 *Of Timber, Iron and Stone* study, updating the history of bridges of the River Lee and doing so in the following framework.

Chapter 1 outlines the early development of the city from its beginnings as a Celtic monastic site through the ravages of the Viking raids to the development of a Norman town. Few, if any, accounts remain of the part bridges played in these phases of Cork's growth. However, it was during these periods that the earliest North and South Gate Bridges were built. These bridges were constructed of timber, but, early in the eighteenth century, stone replaced these timber constructs, resulting in increased security and permanence of structure. Following on from the destruction of the city's walls during the siege of 1690, the city grew eastwards, aided by the construction of a number of small bridges and, in particular, by the building of the city's main bridge, St Patrick's Bridge, in 1789. The story of this bridge and subsequent structures at the same site is told in Chapter 2 of the book.

Chapter 3 deals with the first half of the nineteenth century, a period in which the city continued to expand and bridge-building played an increasingly important role in the civic and economic developments of the time.

During the second half of the nineteenth century, dealt with in Chapter 4, a further replacement of the crucial North Gate Bridge occurred along with two other significant bridge developments. The story of the replacement of Anglesea Bridge and an issue regarding its naming, locates the story of Cork's bridges at the heart of the nationalist endeavours of that period.

The arrival of twentieth century saw the transport revolution of the previous fifty years manifest itself on the streets of Cork with trams, railways, bicycles and

Water Gate of the medieval town, where Castle Street is today.

motorcars becoming part of everyday life. It also saw an escalation of violence in the pursuit of independence, all with consequent effects on the bridges of the city. This period is dealt with in Chapter 5 of the book. The 1980s can be said to have truly been a decade of bridge-building in the city with no fewer than four new river crossings built, also recounted in Chapter 5. By this point, there were some twenty-seven bridges spanning the North and South Channels of the Lee and, during the remainder of the twentieth century and the first decade of the twenty-first, some of these would disappear forever while a further five new bridges were built. These are dealt with in Chapter 6.

Chapter 7, the final chapter, takes the form of a short descriptive tour of the river channels, starting from the west of the city and travelling downstream on the South Channel, before turning where the channels rejoin to the east of the city centre and journeying back upstream on the North Channel.

The story told in these pages may not be that of the world's greatest bridges but it demonstrates that, where bridges stand, human endeavour has overcome nature's obstacles and social and economic development has occurred; where bridges stand, an account of the past is written in the stones and arches that adorn the place; where bridges stand, people with vision and power have made their mark and have left to future generations a legacy to which they can add their own history.

1

EARLIEST CORK

There are few references to bridges. Those that existed seem to have been comparatively flimsy timber structures (perhaps pontoons), easily destroyed or else a sort of openwork causeway – an elaborate form of stepping stones.[9]

A Celtic Beginning

The history of Cork City is said to begin around the year AD 606, with the foundation of a monastery on a hill overlooking a wide and marshy valley, where the headwaters of the Lee spread around a number of small islands, as the river flows eastwards towards the sea. Fionn Bairre, the patron of this monastic settlement, is said to have journeyed from his monastery in the remote valley of Gougane, some 40 miles to the west. From there, he followed the course of the River Lee that rose at Gougane, until he came to the valley, a place known as Corcach Mór Mumhan, the Great Marsh of Munster.

One of the earliest references to the monastic settlement at Cork is found in the Annals of the Four Masters where it is recorded that Suibhne, son of Maelumha, successor to Bairre of Cork, died in the year 680.[10] Over the following centuries, several hundred holy people came to this place of prayer and learning, among them Coleman, Nessan and Garvan, who gave his name to one of Cork's earliest suburbs, that of Dungarvan, a place no longer appearing on the maps of Cork, but believed to have been in the North Main Street area of today's city.

With these holy people came the ordinary folk, the Manaig. It was usual at this time for people to gravitate towards monastic settlements, where the spiritual leader took responsibility for both the monastery itself and the lay society associated with it. This society consisted of pilgrims, penitents and the tenants who cared for the monastic lands along with the labourers and farmers, craftsmen and tradesmen.[11] The monks provided spiritual and educational services to these people in return for produce and labour.[12]

This Celtic period in Cork's history was to last until the ninth century and the coming of the Vikings, the people from the north.

Viking Period

Viking attacks on Ireland occurred in two phases. They first attacked Cork in the year 820, following which the settlement was plundered on a number of occasions. Across the country, from about 850, 'the great raids of the ninth century were over'.[13] The second phase of penetration began early in the tenth century and this time many of the raiders who came stayed. Their assimilation into Irish society over the ensuing century can be sketched in broad strokes. Throughout the country, groups formed alliances with local kings and were, 'drawn more and more into Irish affairs, playing their own parts in the complex and shifting alliances of the little kingdoms'.[14] Others formed bases in harbours around the coastline and developed centres of trade. Thus, in time, the Norse settlements of Waterford, Wexford, Cork, Limerick and Dublin, among others, developed and, 'the population of the Norse towns turned Christian and finally in speech and habit almost Irish.'[15]

In the Great Marsh of Munster, generally believed to have been where the North and South Main Street areas of today's city, a new centre of population developed and Cork changed from being a monastic settlement into a town. In particular these Norsemen were thought to have settled on the mainland where St Nicholas' church stands today, and then moved across the river onto the southern island of the old town. If this is so, then the earliest regular crossings of the river will have been done by these Vikings, probably at the site of the South Gate Bridge, the oldest bridge standing today. Recent excavations have uncovered a, 'wooden walkway extending from the south bank of the river and conceivably across it to the south island, very close to the South Gate Bridge'. Just downstream was, 'the little wooden harbour ... where small ships and boats put in' and this walkway was used by Cork's Hiberno-Viking merchants engaged in trade.'[16]

Modern study generally associates the Vikings more with trade than with mayhem, although the traditional violent image of the Norsemen still prevails at times. Violent or not, the Vikings did become part of Cork's make-up and the Viking period in the city's history lasted well into the eleventh century.

The Normans

After 1169, Cork was to succumb to invasion yet again, this time by the Normans. The arrival of the Normans brought to Cork the third element in what Bradley and Halpin called the, 'fusion of many diverse cultural traditions' contributing to the emerging culture in Cork.[17] Such was the importance of Cork that Diarmuid McCarthy, King of Desmond, had chosen it as his capital in the twelfth century. When the Normans proved too strong a foe, he finally surrendered the town to them in 1177 and an English governor and garrison were introduced for the first time. England's hold over Cork was consolidated by King Henry II granting the surrounding lands to his knights Robert Fitzstephen and Miles DeCogan and then subsequently through the granting of charters, the first of which was given by John, Lord of Ireland, prior to 1189.

The chartered borough was one of the standard methods of economic development employed throughout medieval Europe.[18] Cork's charters were a royal declaration of administrative government; they also encouraged investment and economic development and marked the participation of Cork in this European movement. The city's second charter, granted in 1242, encouraged merchants to acquire holdings in the city by granting them freedom from tolls and customs as well as a monopoly of trade within their own jurisdiction.[19] From this time, the constructional development of the city followed the pattern of most Norman towns throughout Britain. Thick stone walls were built for protection and two drawbridges were built at the northern and southern entrances to the walled town. As well as this, another bridge joined the town's two main islands, commonly known as the Middle Bridge.

Thus, several hundred years after the arrival of the earliest inhabitants and the founding of the settlement, Cork acquired the formal shape of most other cities and towns and, being an island settlement, solid bridges for communication and connection with the mainland.

The Walled City of Cork

The stone walls, built by the Normans, were known as the King's Walls and ran downstream from the North Gate Bridge to Corn Market Street, southwards along Corn Market Street and the Grand Parade to the South Channel of the river, westwards upstream to the South Gate Bridge and contunued up to the site of the present Clarke's Bridge, northwards along Grattan Street until they reached the North Channel of the river again, and finally eastwards back to the North Gate Bridge. Some twenty watchtowers were built on the walls and the entire fortress was to remain intact for about 450 years.

Within the walled city were castles, such as those located at the eastern end of Castle Street (the King's and Queen's), and Skiddy's near the northern gateway. There were churches such as St Peter's and Christ church, a strong gaol and a mint. Beyond the walls, families such as the Barrys and the Gallways lived in their castles. Following the Norman invasion, religious orders settled in the land beyond Cork's walls and they contributed to the development of the suburbs around the town. In 1214, the Grey Friars (Franciscans) founded a monastery at the foot of the northern hills and this was followed in 1229 by the founding of a Dominican Friary at St Mary's of the

Isle. The Augustinians built their church where the Red Abbey Tower still stands today and the development of such habitations beyond the enclosed city added to the importance of the bridges in the life of Cork.

In the following centuries, the fortunes of Cork were directly associated with that of the surrounding hinterland. When, as in the fourteenth century, famine, disease, climate change, and war mongering caused agricultural productivity in the region to decline, the effect on the city was calamitous, as the area, 'could no longer provide sufficient food to feed the people of the city.'[20]

The riches of those who dwelt within the walled city were a constant source of temptation for some of those less well off who lived outside. These outsiders would often attack the city, leading to considerable damage being inflicted on the walls and also on the bridges. In 1319, the Norman rulers from afar presented their Cork brethren with grants for constructional development, including the building of bridges. Hostilities continued and, in 1376, the King allowed remission of monies due to him as, 'the walls of Cork, then stated to be in a state of great dilapidation, being by the hostile incursions of the Irish enemy, almost totally destroyed.'[21]

The earliest known map of Cork comes from Tuckey's *Cork Remembrancer* and dates from 1545. This is said to have been copied from a sketch in the Tower of London and shows fifteen islands, two of which form the main city and a third to the southwest called Abbey Isle, the site of the Dominican Friary. The map shows three bridges in all, the North and South Gate Bridges and a third joining the north and south islands of the city, the aforementioned Middle Bridge.

A more detailed map of Cork (though in truth it could be called a sketch) dates to 1590. It is titled *Walled City of Cork circa 1590* and clearly shows the watchtowers, a water-gate (located at the Daunt Square end of Castle Street) through which small ships or barges could enter the city, and the abbeys and castles outside the walls. Again the three bridges of the 1545 map are visible and it is interesting that this sketch format gives an idea of what the bridges actually looked like. Those at the north and south gates appear to be of timber construction, with supports driven into the river bed at the North Channel. The Middle Bridge is an arched construction and it is evident that consideration had been given to its appearance at its location in the centre of the town.

Speed's map of 1610 shows the beginnings of movement beyond the city walls, in particular to the northeast or Paul Street area of today. Note numbers 13 and 14 on the map legend, an entrance fort and a walkabout.

The Siege of Cork

The seventeenth century was a time of great turmoil for the people of Cork. A rebellion in support of Charles I against his adversaries in Parliament began in Ulster in 1641. It spread southwards through Munster and resulted in the expulsion of the Catholic population from the City of Cork on 26 July 1644,[22] on which day also, 'the civil authority ceased in Cork'.[23] By the end of the 1680s, 'the real estate and resources of Cork City and County had been transferred into the possession of the Protestant English community.'[24] Civic administration was completely in Protestant hands; a census

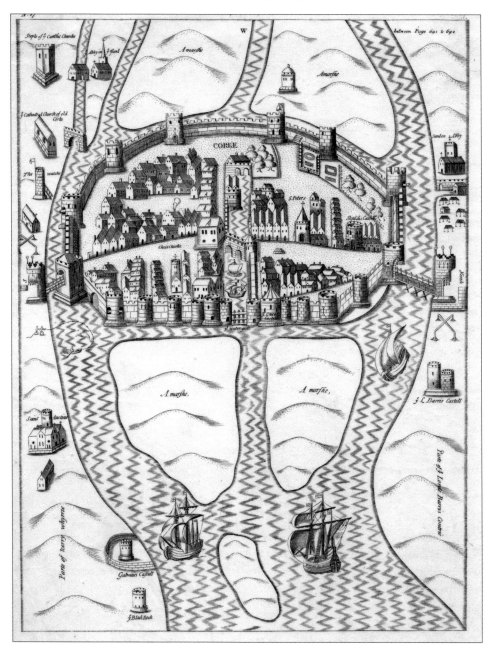

Sixteenth-century map of the Pacata Hibernia. (Courtesy of Cork City Library.)

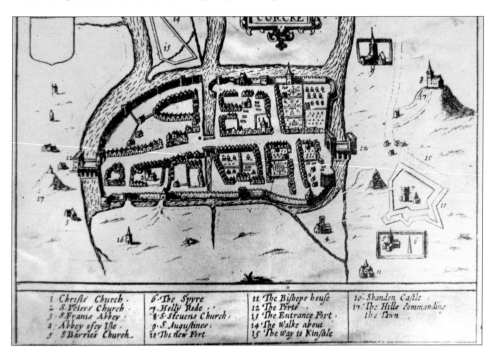

Speed's Map 1610. (Courtesy of Cork City Library.)

of Ireland in 1659 showed that almost two-thirds of the Cork City population were categorised as English.[25]

The supremacy of the Protestant population within Cork City was challenged in 1686, when the Catholic James II ordered that his co-religionists have their status restored. When the Protestants failed to comply, he reconstituted the corporation, giving the Catholics a two-thirds majority. This was a short-lived victory for the Catholics, however, as, in the war between James II and William of Orange in 1689/90, Cork was besieged and the city was taken for King William by the Dukes of Marlborough and Wirtenberg after less than a week's resistance.[26] It was during this siege that the walls were breached for the first time. Cannon mounted on the Red Abbey Tower bombarded the walls near to where the city library is today.

Bridges too had a role in this five-day siege, for although the main north and south crossings were drawn up, it was across the small drawbridge between Castle Street and the Paul Street Marsh that Colonel Munchgaar and Wirtemberg's troops forced the retreat of some 200 defenders of the city who had occupied this outlying part of Cork.

The route of the advancing troops and this drawbridge can clearly be seen in a map of the siege preserved among Marlborough's papers and reproduced in the 1990 *Journal of the Cork Historical and Archaeological Society* article by Diarmuid Ó Murchadha. This drawbridge, however, is not visible in yet another map from Tuckey's

Cork Remembrancer dating from 1690, the year of the siege. This 1690 map shows a city not unlike that of a hundred years previously, though suburbs have grown up to the north and south, and a bowling-green is seen to the east of the walls. Only two principal bridges remain at the northern and southern entrances of the town. The siege of Cork and the resultant breaching of the walls was the catalyst that led to the expansion of the city over the next 100 years.

Some key economic events also occurred in these years, which greatly encouraged Cork's expansion. In 1608, James I had granted a charter to the city, on which it was stated, as privilege number four, 'our aforesaid City of Cork and all ... standing, lying, being and extending from the outside part of the wall of the said City of Cork on all sides whatsoever for the space and circuit of three miles ... shall be one county of itself ...'[27] The cattle acts of the 1660s, preventing the import of Irish cattle into Britain, forced producers to seek new markets, which they successfully did through increased slaughtering in Cork City and exports to the expanding transatlantic British colonies and the Caribbean. The quality and relative cheapness of the Cork produce drew large fleets of merchant ships to the port, resulting in more employment for large numbers of people who now migrated to the city. These were major contributing factors to the most significant development during this period which would continue throughout the eighteenth century, the reclamation of the marshlands beyond the old city walls. In 1692, the people of Cork petitioned the Lord Lieutenant for permission to open gates in the walls and develop the lands beyond. Permission was granted and the corporation leased the marshlands to willing developers, and men such as Alderman Dunscombe and Tuckey obtained possession of entire islands on which they developed streets, quaysides, and warehousing.[28] In time, many of the islands were joined together and a number of the river's channels were bridged to achieve this.

The eighteenth century saw a continuation of this development of the new city. A number of the channels of the river that ran between the islands were covered over and new streets were created. St Patrick's Street, for example, is a curved thoroughfare because at one time it was such a river channel. To allow shipping to enter and leave that channel, a drawbridge, from today's Drawbridge Street to the Merchant's Quay shopping centre, spanned the waterway. The development of the street, which began in 1783, meant that the drawbridge had to be removed and, in 1787, a notice appeared in the press stating that:

> ... the present state of the Old Drawbridge has by several of the inhabitants in the city been represented to me as in a very decayed and ruinous condition and most dangerous to passengers and as it now ceases to be of material utility on account of the arching in Patrick Street to nearly approaching it, I, therefore, at the desire of several of the citizens and for the general safety of the public, give notice that the timbers, irons etc. belonging to said bridge will be sold by auction ...[29]

Clarke's Bridge, 1776. (Courtesy sandraocallaghan.com.)

Cork City old and new. Clarke's Bridge, 1776. (Courtesy sandraocallaghan.com.)

At Daunt Square another bridge crossed the river channel while further along, Tuckey's Bridge spanned from Tuckey's Street to today's Oliver Plunkett Street. The first bridge between these sites was likely to have been that built during 1698/9 by agreement between Timothy Tuckey and William Dunscombe, 'whereby the latter was to build a stone bridge from Tuckey's Kea [*sic*] to the east or Dunscombe's Marsh'.[30] The other bridge was that built in 1705, linking Tuckey's Quay with George's Street (Oliver Plunkett Street today). So once again it can be said that the building of these bridges helped the growth of the City of Cork.

Some of these bridges appear in eighteenth-century maps of Cork, maps that well illustrate the city's development.[31] Despite the significance of these structures, the most important bridges were, as they had been for centuries, those at the North and South Gates of the city.

The North and South Gate Bridges

During the seventeenth century, these two bridges were subjected to alterations because of the river they spanned. In the early 1630s, there was, 'a prodigious flood in the River Lee which, among other damages done to the City of Cork, carried away both the North and South Bridges ...'[32] In 1638, at the council's behest, a new structure was built at the North Gate[33] and, in 1676, the South Bridge was rebuilt.[34] Following damage again caused by flooding in 1678, Lord Shannon, governor of the city, had new drawbridges built at both sites.[35]

Many of these endeavours appear in the minutes of the corporation as recorded by Richard Caulfield, who, in some cases, named the individuals involved. When, on 3 February 1709, it was yet again suggested that these timber bridges be replaced, he recorded that it was stated that, 'it would be very convenient that there should be a stoney bridge if it could safely be done' at the North Gate. Present were Mr Mayor, both Sheriffs, Aldermen Rogers, Crofts, Crone, Chartres, French, Roberts, Whiting, Andrews, Cotrel, Delahoyd, Mr Walker, Mr Perrie and Mr Browne.[36]

Two years later, on 7 May 1711, agreement was reached that such a stone bridge would replace the existing wooden structure at the site and that Aldermen Crone, Goddard, Delahoyd, and Chatres, along with Mr Richard Phillips, would oversee the operation. Goddard was appointed treasurer and would agree the cost with the stonecutters at a suitable rate per foot of stone. A year later, in April 1712, it was agreed that, '100*li* be taken out of the iron chest and paid to Ald. Will Goddard towards building the North Bridge'. On 12 January 1713, 'forasmuch as the South Bridge of this City is in a dangerous and tottering condition, agreed that the same be built with stone and that Alderman Chatres and Goddard, Mr Phillip French, Mr Will Lambley and Thomas Tuckey be over-seers of said work ...'[37] Further information was forthcoming the following week when, at a meeting on 18 January 1713, it was recorded that:

The persons appointed to oversee the building of the new South stone bridge reported that they were upon agreement with Thomas Chatterton, mason and John Coltsman, stone-cutter, to build the same by the great to wit, 300*li*, and the old wooden bridge, the corporation finding what cramps may be thought fit and allow them to use of what centres may be convenient, with boards for the same, also to give them all the tarras left of the North Bridge, with some other small privileges, which agreement was approved.[38]

It therefore appears that work on the North Gate Bridge started before that on the South Gate. Tuckey wrote that in 1712, 'the wooden bridge at the north end of the City of Cork was taken down and a new bridge erected, the piers, arches and abutments of which were faced with hewn stone.' He went on to write that in 1713, the wooden bridge at the south end of the City of Cork was taken down and a stone bridge erected, 'at the expense of the corporation'.[39] The South Gate Bridge of 1713 is the same one that serves the city to this day, although it was widened to the specifications of Alexander Deane in 1824.[40] The original was approximately 15ft wide with two piers of 4ft 5ins. The centre arch was 26ft with the side arches being 21ft and 23ft respectively.[41] Both the North and South Gate Bridges of the early

South Gate Bridge in the eighteenth century.

eighteenth century were similar in appearance, having elliptical arches and faced with hewn limestone. The North Gate Bridge was overhauled in 1831, replaced again in 1865 and yet again in 1961, as we shall see later.

The reconstruction of the North and South Gate Bridges in stone, rather than timber, was a further statement of the permanence of the expansions that had taken place beyond the old walled town of the late medieval period. The bridges were faced with hewn limestone which combined the practicality of weather protection with the desire to be aesthetically pleasing.

Other Bridges

Other bridges built prior to and about this time are seen in Carty's map of 1726. They include the crossing at Daunt Square and Tuckey's Bridge joining Tuckey's Street to today's Oliver Plunkett Street (then George's Street), across the channel that ran where the Grand Parade is today. This map, incidentally, was printed in one of Cork's narrow lanes, Cockpitt Lane. (Close examination of this map shows that Carty, the map-maker, had as his business logo an image of two fighting cocks and the reason that Cockpit lane was so named is because it was the location for cock fights in the city.)[42]

Tuckey's *Remembrancer* tells of yet more bridges; in 1728, 'a wooden bridge was built on Dunscombe's Marsh by Alderman Crone'. (Dunscombe's Marsh included the area where the English Market is today and through which Oliver Plunkett Street runs.) In 1731, 'a bridge on Hammond's Marsh which led to the Quaker Meetinghouse' was built. (Hammond's Marsh was located to the west of today's Grattan Street.) In 1732, 'a large bridge was erected between Hammond's and Pike's Marshes'. (Pike's Marsh was located to the western end of Batchelor's Quay.) 'A new springing bridge was erected where the old drawbridge was in 1772; on 29 January 1773, a child was found near Peter's Church Bridge; on 2 March 1775, the public was cautioned in a Cork paper to 'be careful in passing at night from Broad-Lane to Fishamble-Lane through Cross Street as the slip near the little bridge was quite out of repair ...'[43]

Reilly's Bridge, linking Reilly's Marsh (or the Distillery of today) with the North Mall and more commonly known today as Wyse's Bridge, was also in place at this time. In a map entitled 'A Plan of the City of Cork in the Year 1750' from Smith's *History of Cork*, we see a much expanded city with many of the bridges mentioned above included. The North and South Gate Bridges are still the principle river crossings, and, in fact, the only ones on the main river channels. Amongst the other bridges evident are that which connected Castle Street to the Grand Parade at Daunt's Square, Tuckey's Bridge and the drawbridge where the Fr Mathew Statue is today, crossing the channel that ran where St Patrick's Street is now and which gave rise to the name Drawbridge Street. As well

South Gate Bridge with Proby's Bridge and St Finbarr's Cathedral in the background.

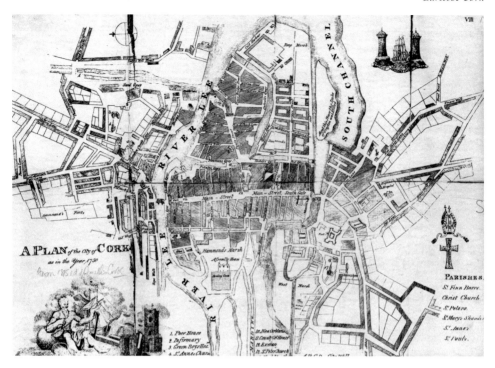

1750 map from Smith's *History of Cork* showing an expanded city. (Courtesy Cork City Library.)

as these, a number of bridges also crossed the channel that ran where Grattan Street is today, to the newly developing lands of Hammond's Marsh.

Samuel Hobbs Bridges

As the city grew and developed with the reclamation of the marshlands, the need for more river crossings became apparent. In 1761, an act was passed providing for, 'the building of a stone bridge not exceeding in breadth 26ft from the quay opposite Prince's Street in the City of Cork to Lavitt's island and a stone bridge from thence to the Red Abbey Marsh, of the same breadth, with a drawbridge in its centre.'[44] Prior to this, the matter had been discussed at corporation meetings. On more than one occasion it had been the view of the councillors that consent would be granted providing those undertaking the project did so at their own expense and gave assurances that the bridges be kept in a state of constant repair without any charge on the corporation. The site of this bridge is not easily recognisable, but the 'quay opposite Prince's Street' referred to is today the southern side of the South Mall. The latter bridge referred to is the site of today's Parliament Bridge. The first Parliament

Bridge at this site was completed and opened to the public on 22 September 1764, the wooden centre being the last section finished.

The occasion of the opening was also that of the anniversary of 'their majesties coronation' and so troops on duty, 'marched thither and fired three volleys in honour of the day'.[45] This was one of the many occasions in the annals of Cork's bridges where opening ceremonies were associated with anniversaries and other significant events in the governance of the ruling order of the day. By association then, the importance of the occasion (in this case the anniversary of the coronation) and that of the structure being opened (the first Parliament Bridge) were connected and the rituals engaged in served to stress this fact.

Two names have been associated with the building of the first Parliament Bridge: Samuel Hobbs and George Randall. There is also a suggestion that they collaborated on the development of the stone and timber sections of the river crossing.[46]

Samuel Hobbs was also involved in the construction of Wandesford's Bridge fifteen years later, in 1776. This bridge was begun in August of that year and Samuel Hobbs superintended the entire construction, which connected Wandesford's Quay, Clarke's Marsh and Crosses Green with the city. In September, a notice appeared offering a commodious dwelling house and stable near the new West Bridge and fronting Crosses Green.[47] Although it is known today by the other geographical term associated with its location – Clarke's Bridge – for many years it continued to be called Wandesford's

Reflections beneath Wyse's Bridge.

Bridge. In 1806, for example, the *Cork Mercantile Chronicle* advertised a comfortable dwelling house at Queen's Place, near Wandesford's Bridge, to be let.[48]

This bridge was the next step in further connecting the island city with the surrounding countryside and, although parts of it had to be rebuilt within ten years of its construction, the bridge has stood the test of time, standing today as one of the oldest bridge structures in the city. Consisting of a single span of 68ft, Peter O'Keefe and Tom Simington have noted that, 'the model for this bridge would appear to be the famous Pontypridd Bridge, built by William Edwards over the river Taff in Wales in the 1750s.'[49] It is hard to believe that today, more than two centuries later, we regularly cross over the same Clarke's Bridge that our forefathers crossed all those years ago.

A Second Bridge for the North Channel

In the Cork of this time, many notices appeared concerning the city's bridges; in 1771 the public were cautioned to, 'be careful how they passed over the old drawbridge after dusk, it being so old and out of repair ...' The drawbridge was finally removed in 1787.[50] As seen previously, on 2 March 1775 a caution was issued regarding the passing from Broad Lane into Fishamble Lane through Cross Street owing to conditions on the little bridge'.[51]

The South Channel of the Lee now had three bridges to cater for the expanding city – South Gate, Parliament and Clarke's Bridges – but the North Channel only had the North Gate Bridge. Naturally then, there was much talk of putting another bridge across the North Channel and, from about 1780, serious planning got underway.

2

St Patrick's Bridge

Walk back now again, by the quays and the river
And the bridges – gan dabht – nobly spannin' the strame,
With the new one (St Patrick's), the finest that ever
Was built, but as yet we have only the name,
Till the Council decides whether wood, stone or iron,
Or brick, the material for buildin' shall be;
When that will be, mo bhrón! we've no bisness inquirin' –
Och, 'Cork is the Eden for you, love, and me.[52]

An Expanding City

Examination of contemporary maps of Cork City shows that it was during the latter half of the eighteenth century that many of the streams dividing the islands on which Cork was developing were bridged over, with a consequent expansion of the city centre. Along with this, some areas on the hills surrounding the city were also settled, among them Dillon's Cross of today, Strand Road - now Lower Road - and other parts of the north-eastern quarter. The residents of these areas were possibly among the first to seek a down-river crossing from the North Gate Bridge, the sole access road to the northern hills. Whether or not they would have been successful on their own will never be known; they were helped in their demands by the fact that the city's commerce was rapidly developing along with the city itself.[53]

Events many thousands of miles from the City of Cork were to have a bearing on commercial developments around this time. The port was used as a rendezvous for shipping going to and coming from the West Indies and the east coast of America, up to and during the American War of Independence. Wool-combing, cloth-weaving, and branches of cotton manufacturing, as well as other industries such as paper-making and glass-blowing were all important to the city during the 1760s and the 1770s. Later, the provision of meat for the navy, along with tanning, brewing and glue-making, were vital industries. However, since the middle of the seventeenth century, the butter trade was the principal source of wealth for many merchants in Cork and butter was exported to every part of the world in firkins.

That many of these industries were based in the north-side of the city was one important factor in the St Patrick's Bridge story. Another was that the river channel following the course of St Patrick's Street was covered over during the 1780s. O'Sullivan's *Economic History of Cork* tells us that the rapidly expanding victualing trade was another factor necessitating a new bridge. The corporation of the day was called upon to build a bridge across the North Channel of the river where the St Patrick's Street stream joined the main channel. This action would bring to completion the project in which St Patrick's Street itself was built.[54]

Opposition to the Project

The idea did not go without opposition. Many businessmen, particularly those in the Mallow Lane (Shandon Street today) and Blarney Lane areas, where the provision and slaughtering industries were concentrated, saw the project as a threat to their own interests. So too did the operators of the many ferries plying the river near the proposed site of the new bridge. They began a campaign to prevent the project going ahead and opposition mounted to such an extent that a public meeting was organised for 18 May 1785 at twelve o'clock at the council chambers. Those called upon to attend were:

> Such persons who have the real interest and welfare of the present city at heart and those whose property will be materially injured by the building of a bridge below the Custom House which will be the cause of depopulating and laying waste the thickly inhabited flourishing parts of the now city ...[55]

One of the principal arguments made was the logic of placing a bridge downstream from the Custom House. (At that time this was located in today's Municipal Art Building on Emmet Place, which had been constructed in 1724.) The organisers of the opposition drew up a petition, calling for the defeat of a project which they claimed would be the ruin of thousands.

Finance Raised

The petition was sent to Parliament in response to corporation representations in favour of the bridge which were made during 1784 and 1785. The opposition was in vain, however, and in 1786 an act was approved for raising the money required for the building of a bridge at a site downstream from the Custom House, where the

Anno Regni Vicesimo Sexto

Georgii III. Regis.

CHAP. XXVIII.

An Act for Building a Bridge over the Northern Channel of the River *Lee*, in the City of *Cork*, and Suburbs thereof, and for other Purposes relative to the said City.

WHEREAS the City of Cork, and the Suburbs thereof, are by the Extension of Commerce considerably encreased, and are likely to encrease: And whereas there is no Communication or Passage from the Country into the said City, on the North Side thereof, for Goods, Cattle, Carriages, or Travellers, except by one Bridge called North-Bridge: Be it therefore Enacted by the King's Most Excellent Majesty, by and with the Advice and Consent of the Lords Spiritual and Temporal, and Commons in this present Parliament assembled, and by the Authority of the same, That the Mayor, Recorder, Aldermen,
* 9 U 2 and

1786 Act authorising the first St Patrick's Bridge.

St Patrick's Street stream joined the main North Channel of the river. The public were given the opportunity to benefit financially from the construction of the bridge. Money for the project would be raised by the issuing of debentures or loan-stock, and the rate of interest was set at 8 per cent. The corporation itself invested £1,000. Tolls were to be placed on the bridge to meet this financial requirement. The tolls were to last for a period of twenty-one years and anything over and above the interest and maintenance costs would be used to pay off the principal owing. There was no shortage of willing investors and the project quickly got underway.

Michael Shanahan was a sculptor and architect working in the city at the time and he was appointed architect and chief contractor on the project.[56] He was an important figure in the architectural world; among his clients were the Earl of Bristol and the Bishop of Derry. *Holden's Directory of 1805* listed Shanahan as working out of White Street.[57] It was during the first half of 1788 that he set about organising the project.

Angry Flood

On 25 July 1788, the foundation stone for the new bridge was laid and over the next six months the bridge began to take shape. The arches reached out from the quay walls and the population marvelled at such an engineering endeavour. Then, disaster struck. On 17 January 1789, a flood, 'such as had not been seen in the city in living memory' swept through the valley, submerging everything in its path in a matter of hours.[58] The entire city, from the Mayor's residence (the present Mercy Hospital) to the lower reaches of the harbour, were completely covered, forming

Level used on foundation stone of St Patrick's Bridge located at Masonic Hall, Cork City.

Painting of the laying of the foundation stone of St Patrick's Bridge.

an inland lake. Only the vigilance of the citizens prevented a major catastrophe. For the new bridge, however, it was a disaster. A boat which had been moored at today's Carroll's Quay (then the Sand Quay), broke her moorings and crashed against the unfinished centre arch of the bridge and destroyed it.[59] Shortly after this, the other arches came down and nothing remained of all the work of the previous year. Michael Shanahan was a ruined man. He could not continue the project and departed for London. There, he met a man named Hargrave and outlined his sorrowful tale. Hargrave took over the contract and, along with Shanahan, returned to Cork and continued the project.[60]

The new bridge was opened on 29 September 1789 and the *Hibernian Chronicle* of 1 October gave the following description of the events as they occurred on the day.

> On Tuesday the Key-stone of the last arch of the new bridge was laid by the Ancient and Honourable Society of Free Masons of this city. The morning was ushered in with the ringing of bells; an immense crowd assembled in the principal streets before the hour of eleven. About twelve the procession of the different Lodges, dressed with their jewels and the insignia of the respective orders, preceded by the band of the 51st regiment, began in the following manner:
>
>
>
> Army Lodge,
> Grand Tyler with drawn sword,
> Grand Almoner bearing a chalice of wine.
> Two Grand Deacons, the Bible supported by two other Grand Deacons,
> The Chaplain of the Grand Lodge,
> Lord Donoughmore, Grand Master of All Ireland,
> Joseph Rogers Esq. Provincial Grand Master of Munster attended by two Grand Wardens, secretary etc.
> Tyler of Lodge No.1.
> Two Deacons of do.
> Master, Wardens Secretary etc. of do.
> After whom followed 14 Lodges with their Masters and Wardens in regular order.
>
>

The lodges referred to were those of the Society of Freemasons who were prominent in the city. The *Hibernian Chronicle* again takes up the story:

> The procession moved from the Council chamber amid the acclamation of the rejoicing multitude, through Castle Street, down the new street called St Patrick's Street, and advanced to the foot of the new bridge, which was decorated on the occasion with the Irish standard, the Union flag, and several other ensigns – here they were saluted with nine cannon, the workmen dressed in white aprons lining each side of the bridge; the procession advanced up to the centre of the

last arch where they were received by the Commissioners and the Architect. The last key-stone which was previously suspended; and which weighed 47 hundred, was then instantly lowered into its berth - and the Bible laid upon a scarlet velvet cushion adorned with tassels and gold fringe was placed upon it – His Lordship, as Grand Master, thereupon, in due form gave three distinct knocks with a mallet; the Commissioners were then called upon to mention the name intended for the new bridge, which being communicated, the Grand Master emptied his chalice of wine upon the key-stone and the Grand Master, in the name of the Ancient and Honourable Fraternity of the Free and Accepted Masons of the Province of Munster, proclaimed it St Patrick's Bridge. The whole body of Masons, composed of upwards of 400 of the most respectable gentlemen of city and county gave a salute three times three which was returned by nine cheers of the populace and the firing of nine cannon. After this the procession marched over the bridge and its portcullis, surveyed them, and were again saluted with nine cannon. They then returned back in the same order to the Council chamber.[61]

St Patrick's Bridge foundation stone in 1859.

St Patrick's Bridge at night. (Courtesy Michael Lenihan.)

St Patrick's Bridge by W.H. Gannett of Augusta, Maine.

These events described in the *Hibernian Chronicle* can be seen as ceremonial, even ritualistic, given the prominent part played by the corporation, the Wide Street Commissioners, and the Masonic Lodges. In recent years, there has been a considerable body of work published on the subject of ritual and ceremony which has emphasised the connection between ritual and power. 'Rituals actually construct power,' wrote Catherine Bell. 'They use symbols and symbolic action to depict a group of people as a coherent and ordered community based on shared values and goals.'[62] David Cannadine wrote that pageants and ceremonials, 'seem to confirm consensus, to disguise conflict and to support both hierarchy and community. Ritual is not the mask of force but is itself a type of power.'[63] In her study of monuments, public spaces, and cultural identity, Nuala Johnson has written of the ongoing project of establishing individual and group identities symbolically coded in public monuments and their attendant ceremonials.[64]

Considering the ceremonial opening of the first St Patrick's Bridge in this context and framework gives new meaning to the events of that day. The procession of the Lodges through the streets to the new bridge was a marking of territory by the elite in society and the lining of the route by the crowds of ordinary people, 'the rejoicing multitude', both gave them an inclusive role in the event and indicated their approval through being there. Furthermore, the invited response from the crowds in the form of 'nine cheers of the populace' was both a means of participation as well as an acclamation of their assent. The powerful members of Cork society, 'upwards of 400 of the most respectable gentlemen of city and county', had demonstrated their authority by marching through the streets before the assembled populace; they had been assented to by the cheering crowd; finally they ceremonially opened and led the way to a new and better future, physically manifest in the form of the new bridge.

The bridge itself was narrow and hump-backed, had three elliptical arches topped with open balustrade, and was constructed of limestone. A gateway for shipping, the portcullis, was located at the northern end of the bridge. Hargrave and Shanahan were the heroes of the day and, for the shareholders who had invested in the debentures, their enterprise soon proved profitable, helped no doubt by the efficient management of the toll-collecting which had passed into their hands.

Tolls Trouble

A large number of vessels passed through the shipping gateway, docking alongside the merchant warehouses on the quays between the new bridge and the North Gate Bridge. These merchants, as well as the people who passed over the bridge, felt much aggrieved at the expense they had to bear in tolls. On the other hand, those who had

Tram and early truck on St Patrick's Bridge. (Courtesy Michael Lenihan.)

Trams on St Patrick's Bridge. (Courtesy Michael Lenihan.)

purchased debentures in the project were in favour of the portcullis and the tolls, upon which depended returns on their investments.[65] These diverse perspectives were illustrated in advertisements in the *Cork Mercantile Chronicle* in August and September 1806. On 13 August, under a heading of 'New Bridge', one advertisement read:

> Such persons as feel interested in keeping the Portcullis on the New Bridge open are earnestly entreated to meet at the Council Chamber at one o'clock tomorrow to consider the best manner of opposing the application now made to shut up the same.[66]

On 10 September:

> The Holders of Bridge Debentures will be paid one year's annuity to the fifth of July last by Samuel Randall Wily, Batchelor's Quay, Cork.[67]

Meanwhile, the tolls for the ensuing year were advertised for sale on 13, 25 and 29 August.[68] Those opposed to the portcullis and the tolls, however, formed a committee, continued their lobbying, and eventually pressurised the shareholders into foregoing any further profits and putting the tolls up for auction. The merchants' committee continued to purchase the tolls for a number of years until eventually they accumulated sufficient funds to pay off the principal owning to the shareholders. In this fashion, the tolls were abolished and the bridge became free to all that used it in the year 1812. The portcullis, however, continued to operate.

St Patrick's Bridge from the Lawrence postcard collection. (Courtesy Michael Lenihan.)

Tolls Authorised by Parliament in 1786
to be collected on St Patrick's Bridge. (26 Geo. III, C.28, Sect. XXVL)

	s.	d.
For every coach, chariot, berlin, chaise, chair or calash drawn by six or more horses	1	1
For every coach, chariot, berlin, chaise, chair or calash drawn by less than six but more than two horses		6½
For every coach, chariot, berlin, chaise, chair or calash drawn by two horses or mules		3
For every wagon, wain, car, cart, or carriage of burthen or other carriage with four wheels drawn by four or more beasts	1	1
For every wagon, wain, car, cart, or carriage of burthen or other carriage with two wheels drawn by more than one horse or other beast		3
For every wagon, wain, car, cart, or carriage of burthen or other carriage with two wheels drawn by one horse or other beast		½
For every horse, mule or ass, laden or unladen and not drawing ...		½
For every drove of oxen, cows or neat cattle – by the score and so in proportion for any greater or lesser number not less than four ...		5
For every score of calves, hogs, sheep or goats ...		½
For every score of lambs ...		2½
For every passenger passing over the bridge – each ... except such person or persons as shall be driven in any coach, chariot, berlin chaise, chair or calash, and the driver or drivers thereof, servant or servants thereof, standing behind the same.[69]		½

Though the merchants were successful in achieving this goal, the days of shipping passing St Patrick's Bridge to berth upstream, were slowly ending. The Custom House, which had operated since 1724, was replaced after 1814 by a new one further downstream.[70] As a result, it was not necessary for shipping to travel as far up the river as previously. Furthermore, there was a general belief in the city that the portcullis section of the bridge was not just a nuisance, but dangerous as well. In March 1812, the Court D'Oyer 100 declared it so and sought for it to be replaced.[71] There was a calling for the closure of the portcullis and the filling in of the stream that ran beneath it. Although this was opposed by the upstream merchants, an Act of Parliament was passed in 1822 allowing this to go ahead. There followed a legal battle and, in 1824, it was finally removed with payment of £1,200 to the affected members of the commercial community.

Proposed Bye-Law

At a Harbour Board meeting held on 2 May 1849, the issue of responsibility for the maintenance of the bridges in the city was discussed. This followed damage inflicted on St Patrick's Bridge when a vessel under tow crashed into one of the parapets:

> Yesterday afternoon, a large brig, the Lucretia, laden with Indian corn, was towed by a tug-steamer into the north-channel. When she came opposite her quay berth, a check-line was got ashore and the tug cast her off. The wind and tide being up the river, the vessel kept her way and the check-line having snapped, she went up until she got foul of the bridge, carrying away over three-fourths at one side of the handsomely cut battlement and doing an amount of damage which astonished all observers. The vessel herself got much injury in bow, stern and bulwarks.[72]

At the meeting, it was stated that there were three public bodies – the Grand Jury, the Harbour Board and the Wide Street Board – but that none of them was responsible for the care of the bridge. A suggestion was made that the repair could be done cheaply with 'a plain rubble wall'. However, the Harbour Master wanted the job done properly and that would cost in the region of £40. Board member Mr Harley, warned that public money being used in this way was a bad habit for public bodies to engage in. He had previously contacted the Admiralty regarding other bridge matters and the officials there had suggested that the corporation was responsible for the bridges. Eventually, Councillor Meagher agreed that £50 would be provided by the board to repair the bridge but that bye-laws should be enacted restricting the operation of towing vessels in order to ensure such an incident could not damage the bridge again. The Harbour Master, however, said that if tow boats were to be restricted from bringing shipping upstream, there could be occasions when strong

St Patrick's Bridge photographed by Ross of Cork. (Courtesy Michael Lenihan.)

Damaged bridge following the flood in November 1853 as depicted in the *London Illustrated News*.

westerly winds would prevent certain vessels from reaching the Custom House. To peals of laughter, Mr Meagher suggested that the westerly winds be made subject to the proposed bye-laws.[73]

Damaged St Patrick's Bridge after flood of 1853.
(Courtesy DeBurca Rare Books.)

It's unlikely that anybody remembered the joke four years later, in November 1853.

Another Catastrophic Flood

In that month, disaster struck again when, following a violent storm, another enormous flood swept through the city. By ten o'clock in the morning of 2 November, Great George's Street (now Washington Street), was completely covered in water. Many were trapped in the upper floors of their homes. A blacksmith was reported to have floated downstream as far as Anglesea Bridge on the South Channel where some sailors threw him a rope and rescued him. Quay walls collapsed with the pressure of the onrushing waters – but amidst it all, one structure, the North Gate Bridge, stood firm. Its huge parapets caused the water on the western side to rise 6ft higher than the downstream side, and, when this great volume of water rushed past the bridge carrying with it large quantities of debris including large trees, it caused great destruction less than a mile away.

St Patrick's Bridge was swept away and with it the lives of between fifteen and twenty people. A small group of men and boys had been leaning over the bridge, fascinated by the swirling waters beneath; they were never seen again. Of four women journeying to the north-side, two were lost.[74] One witness saw a, 'woman and two children tumble down the fearful chasm and instantly perish.' 'A woman named Fitzgerald ... having fallen in, she held for a while by the broken balustrade ... but quickly dislodged, she was borne away.' 'A woman named Roche, who lived near the North Chapel, and three young men, labourers from Killeens ... named Murphy, Callaghan and Finnegan are said to be among the sufferers.' 'Margaret Lynch was a married woman who lived in Brown Street; Catherine Daly resided at Gardiner's Hill.'[75]

The Mayor, John Francis Maguire, away in Dublin on business, heard of the disaster and rushed back to the city to take charge of the situation. Firstly, he increased the number of ferries crossing the river and then arranged a temporary pontoon bridge to accommodate the requirements of the citizens until more permanent plans could be made:

Captain White, harbour master, had five barges moored yesterday at equal distances between Merchant's Quay and Patrick's Quay, close to the eastern side of the bridge, on which he had balks of timber [*sic*] laid, over which a flooring will be placed to make a floating bridge. It is expected it will be open to foot-passengers tomorrow and thus save the public the expense and danger of ferry boats.[76]

John Benson, the city engineer, was a native of County Sligo and a man whose architectural endeavours are to be seen in every part of the city to this day. He was instructed to draw up the plans and by the following day was able to present the corporation with drawings for a timber bridge which he recommended should be placed 50ft away from the existing bridge site to facilitate reconstruction. Estimated construction time for this bridge was a month:

> Sir John Benson has already erected a piling machine on boats at the western side of the bridge and will this day proceed with the driving of piles of the wooden bridge, which it is hoped, may be opened before Christmas to all vehicles and passengers, when the pontoon bridge will be removed.[77]

At a cost of 'about £2,000' and paid for by the Harbour Commissioners, five weeks after the disastrous flood, on 7 December 1853, Benson's Bridge was officially opened to the public by the Mayor at a site to the west of the destroyed structure.[78] (In 1859, when a footbridge was planned for a site between Merchant's Quay and St Patrick's Quay downstream of the main bridge site, a report said that it was in the opposite direction to Benson's Bridge.)[79] Before the end of the year, the corporation took the decision to apply to Parliament for permission to replace St Patrick's Bridge as well as the North Gate Bridge which had contributed largely to its destruction. That permission was granted in 1856 when the Cork Bridge and Waterworks Act was drafted into law, allowing the corporation to borrow money for the replacement of the bridges. Before that, however, a number of serious issues arose regarding what type the replacement bridge should be, what material it should be constructed from and the means of financing the rebuilding.

The Issue of Responsibility

As in 1849, following the November 1853 flood, it wasn't long before the issue of who would pay for the repair of St Patrick's Bridge reared its head. At a harbour board meeting on Wednesday 11 January 1854, a letter from the Attorney General was read which stated that, 'respecting the Wide Street Board and the Grand Jury, the corporation are the proper party to repair or reconstruct St Patrick's Bridge'. He also advised that, 'the Harbour Commissioners not interfere' but that, 'if in constructing a new bridge or new arches, a plan could be adopted which would improve the navigation, the Harbour Commissioners may fairly defray any additional expenses that arise therein.'[80]

A week later, the corporation appointed a committee to prepare an act for the replacement of the bridge, while a report on the cause of the destruction was presented to the Harbour Commissioners by Henry Hennessy, MRIA, librarian at Queen's College, Cork. In the report, he stated that the quantity of rain registered at the college for the fortnight before the flood was 9.86ins compared with 3.87ins for the same period the previous year. This gave rise to over two billion extra cubic feet of water in the river. Henry Hennessy concluded that the primary causes of the destruction were excessive and continued rainfall, which resulted in the saturated condition of the basin of the Lee for some time before the flood and the sudden and extraordinary fall of rain during the night immediately before the flood, whereby an immense volume of water was superimposed on the already saturated catchment basin of the river. Significantly, the report also said that swelling produced by bridges and weirs, among which was the North Gate Bridge, was a secondary cause.[81]

An editorial in the *Cork Examiner* of 28 January brought to the attention of the public that in order to finance the rebuilding of St Patrick's Bridge, the bridge committee proposed:

> ... an application to Government for the continuance of the existing tax on coal entering the port, which tax is not higher than five pence on each ton ... By consenting to continue it, they [the members of the government], will themselves be free from applications for grants, and the citizens of Cork will be saved a world of anxiety as to where all the money is to come from.

The editorial also agreed with members of the Harbour Commissioners who believed that serious consideration should be given to making the replacement a swivel bridge.[82]

Over the following month, the coal tax proposal became complicated. Not all imported coal was landed within the borough of Cork, some was put ashore in Queenstown and Ballinacurra near Middleton. The authorities in these locations decided that if Cork was to seek a continuance of the tax for the improvement of the city, they would seek to obtain a percentage for the benefit of their own localities. At a meeting of the corporation on 15 March, the high sheriff said that such was the extent of the claim from a number of places outside of the city, that, 'they would put a gloss on the transaction that is not altogether correct'. With these claims lodged in opposition to the city receiving all of the tax, 'it would be idle to go on'. He therefore announced that he would move at the next meeting that no further steps be taken in this matter'. Furthermore, he said, 'Benson's Bridge is a very good one and will certainly last twenty-five years.'[83] So ended the coal tax proposal, but the swivel one continued for some time longer.

The Bridge Type Controversy

What type should the new St Patrick's Bridge be? That was the question occupying the minds of the councillors and the citizens. Ideas and suggestions were many and varied, with some people suggesting that since Benson's timber bridge would last for many decades, the money could be used for building completely new bridges and the next generation could worry about replacing St Patrick's Bridge.[84]

St Patrick's Bridge, however, was always going to be replaced and the main decisions to be made were whether the bridge should be constructed of timber, iron or stone and whether it should be a swivel bridge or not. Benson, therefore, prepared plans for all three types of bridge.[85] Many arguments were made in favour of and against the different types, but in the end a stone bridge was initially decided upon because there was plenty of limestone available locally and employment would be given to the local stone-cutters and stone-masons.[86] (In fact almost 100 of these people were employed on the project.) Were an iron bridge to be built, it would be more costly and most of the money would leave the area. A fixed rather than a swivel bridge was selected and one of the arguments that John Benson made in favour of this was that, in the event of armed insurrection, the mechanism could be jammed by the rebels, thereby preventing troops from the barracks gaining access to the city via the shortest possible route.

Meanwhile, the ruin of the first St Patrick's Bridge was a cause for concern. At a Harbour Board meeting on Wednesday 3 May 1854, the chairman, Mr Donegan, pointed out that St Patrick's Bridge was falling 'day after day into dilapidation'.[87] On Wednesday 10 May, an editorial in the *Cork Examiner* said that, 'it is high time that the vacillation was at an end and that some plans were determined.'[88] Following this, the corporation asked John Benson and another engineer, Mr J. Long, to do a full report for them. The inspections for this took place on Tuesday and Wednesday 11 and 12 May, following which John Benson recommended that, 'the present wreck should be taken down and that a flat bridge, with a spacious archway should be constructed in its stead.'[89] Long, in his report, largely agreed with Benson. When the report was read to the corporation, Alderman Lyons said that, 'whatever discrepancy there might appear between the reports, there was in reality no difference'. He further pointed out that while specifications and plans would be prepared, 'there would no doubt be a war about swivels.'[90] In a supplementary report to the corporation dated 31 July, Benson further recommended that, 'the injured piers be entirely rebuilt on a new foundation several feet below the present deepest bed of the river.'

It took some time for the plans to be drawn and in April 1855 it was proposed to a joint Harbour Board and corporation committee that, 'it would be advisable to have erected on the site of St Patrick's Bridge a solid bridge as level as possible, width 64ft and a deeper foundation than the former one.'[91] The same notice was read to the corporation the following Monday, but the issue became acrimonious. Some members still agitated for a swivel bridge and when pressed, one of them, Mr Fitzgibbon, acknowledged holding property above the bridge. Mr Lambkin declared himself, 'perfectly disgusted with the personal motives which people bring into every public question. It is really monstrous.'[92] The end result was that the issue was postponed until the next meeting.

Reporting on the matter, the *Cork Examiner* did not deal with the personalities involved, but came firmly down on the side of a swivel bridge. On 4 May, it declared that, 'the proposal to replace St Patrick's Bridge with a swivel rather than a fixed structure affords perhaps the greatest opportunity which has ever offered for improving the City of Cork.'[93]

The issue remained dominant for the summer months, until September, when at a corporation meeting, not only was the swivel bridge again proposed and rejected, but there was another proposal to finance the bridge through a coal tax, exclusive of the coals landed in those areas that had previously opposed the plan. This prompted a letter to be published on 1 October saying that the proposal was, 'pregnant with mischief' and that in opting for a solid structure the corporation was, 'highly unjust towards the owners of property situated contiguous to the quays above St Patrick's Bridge and the northwest portion of the city generally.'[94]

Such was the pressure in favour of a swivel bridge that the matter was again discussed and voted on at a corporation meeting held on 12 November. The result was twenty-one votes to eighteen in favour of the solid structure. On 14 November, the *Cork Examiner,* arguing passionately in favour of the interests upstream of St Patrick's Bridge, declared that this decision could not be, in the opinion of the vast majority of the citizens, 'a more unwise or impolitic decision'. On 30 November, the editor continued his campaign against the corporation decision saying that, 'it is not likely that the decision of the council on the question of the bridge will be taken as binding on the citizens and the preliminary steps have already been taken for expostulation and remonstrance ... perhaps opposition ...'[95]

On Monday 3 December, a huge public meeting took place regarding the issue. It took place in the Court House at two o'clock and was at times acrimonious. When the Mayor, Sir John Gordon, attempted to direct proceedings in favour of the corporation's position, John Francis Maguire (the owner of the *Cork Examiner*), to cheers from his supporters, proposed that, 'this meeting is deeply impressed with the importance of throwing open the navigation of the North Channel by means of a swivel bridge to replace St Patrick's Bridge ...' The resolution contained a suggestion that the swivel need only be opened by night, thereby minimising disruption to traffic.

The debate continued, dealing with matters such as the annual maintenance cost of a swivel bridge versus a solid one and whether in fact two swivels should be employed in order to minimise disruption. Mr Edward Scott proposed that the proceedings of the previous corporation meeting, at which a solid structure was decided upon, be endorsed. However, Mayor elect for the up-coming term, William Fitzgibbon, said that, 'the people of Cork would not be led away by persons who came there to disturb the great object of their meeting'. He then proposed, 'that the meeting is resolved to oppose in Parliament that part of the bill now sought by the Town Council which proposed to destroy the navigation of the upper part of the north branch of the River Lee by erecting a solid bridge to replace St Patrick's Bridge'. Mr Dowden seconded the resolution and, following some discussion on the financial aspects of the plan, this resolution was carried. A further resolution by Mr Seymour called that the attention of the Admiralty be drawn to these proceedings.[96] The *Cork Examiner* reported glowingly on the meeting's position, declaring that Mayor Gordon had, 'a self-sufficient and rather arrogant tone'.[97]

But this had only been a public meeting and, as such, had no formal standing. It remained to be seen whether the wishes of the majority at the public meeting would be acceded to. On Friday 21 December, details were published of the bill which would go to Parliament allowing for the rebuilding of St Patrick's Bridge and the North Gate Bridge as well as the purchase of the waterworks for the city. It proposed a coal tax of four pence on coal landed within the borough, against which money for the project would be borrowed.

The answer to the public meeting came in the early months of 1856 and saw the corporation still refusing to change its opinion on the issue of bridge type, adamant that it would be a solid structure. Before the bill was debated in the House of Commons, however, letters appeared in the press still calling for a swivel bridge.[98] Also, reservations about the method of finance led to discussions at corporation meetings regarding the coal tax, during which concerns were expressed regarding the effect the tax would have on the citizens and on the trade into and out of the city. As a result, the tax proposal was withdrawn yet again in March.

When the bill was finalised in May the question of bridge type was finally resolved.[99] It would be a solid structure and although calls against this decision persisted, it was not reversed. Permission was obtained under the bill for monies to be borrowed for the project as well as that of the North Gate Bridge reconstruction. Two further issues now came to the fore: calls to actually do the job began to appear in the press and a debate began concerning what the bridge should be constructed from, iron or stone.

Following letters to the press in support of stone by Thomas Jennings, on 29 March 1858, Alex Crichton wrote a letter to the *Cork Examiner* in which he proposed to, 'confute Mr Jennings arguments and to prove by facts that of all the materials generally used in the construction of bridges, wrought iron is the strongest and best suited for that purpose'.[100] In response, on 9 April 1858, a statement was issued by the Society of Stone-Cutters in the City of Cork. In it, as well as thanking their main supporters, they said that:

... we respectfully submit that the corporation should pause before they decide against us – before they expend a large sum of money amongst strangers instead of in our own city – before they seek materials in another country when better materials are to be had in their own valley and most of them on the very spot (in the remains of the old bridge) and before they choose a flimsy costly ugly iron structure which will not with any amount of tinkering and painting last fifty years when, at very little if any more cost, they can have a permanent inexpensive handsome stone structure which will last for ages.

The statement was signed Roger Mathew, president and Michael Walsh, secretary and was published in the local press.[101]

The statement was followed by a general meeting of masons and builders, held on Monday 19 April in Mary Street in the city. A number of resolutions were passed including, 'that the meeting fully agree with the opinions expressed by so many of the enlightened and intelligent burgesses of this city, regarding the greater permanency, desirability and economy of stone in comparison with iron, as a material for the construction of St Patrick's Bridge'.[102] The debate continued throughout the year but eventually John Benson's plans for a three-arch stone structure were adopted.

On Friday 8 and Monday 11 April 1859, the bridge committee met to receive and open tenders for the project. Five tenders were submitted: John Edwards at £17,780; William Joyce of Queenstown at £18,250; Joshua Hargrave Jnr at £14,450; Joseph Enright of Dublin at £11,642 2s and John Moore at £16,350. As Enright's was the lowest, it was decided to recommend it to the main corporation and he was awarded the contract.

By June, however, problems had arisen. An editorial in the *Cork Examiner* on Friday 17 June stated that, 'we hear of the contractor coming before the committee ... with a modest demand to be paid for the work he had already done and to be eased of the burden of his contract'.[103] In July, John Benson was mandated to agree with Joshua Hargrave for the completion of the bridge. By August Mr Barnard had been appointed clerk of works, but this became an issue when, at an Improvement Department meeting in August, Mr Keane wondered why John Benson did not perform this role as well as that of design engineer.[104] The majority of the members, however, approved of the necessity of Barnard in the role and on 11 November 1859, six years after the first St Patrick's Bridge was destroyed by a flood, the foundation stone for the new St Patrick's Bridge was laid by the Earl of Carlisle, Lord Lieutenant of Ireland.

Having lowered the stone into place, he proclaimed to the assembled crowds, 'I declare the foundation stone of St Patrick's Bridge to be duly and truly laid'.[105] There followed three cheers for the bridge, three for the earl and a further three for the designer John Benson. The ceremony proceeded with the conferring of a Knighthood on the Mayor John Arnott.[106] (Under the alias Timothy Tightboots, he was well known for his generosity to the poor people of the city.)

Returning again to recent studies on ritual and ceremony, Nuala Johnson, in her work, refers to Paul Connerton's suggestion that, 'national elites have invented rituals that claim continuity with an appropriate historic past'. On the occasion of the laying of the foundation stone for the new St Patrick's Bridge in November 1859, there were many echoes of the ceremony held seventy years earlier for the first bridge, which again sought to symbolically underpin the place of the elite in Cork society.

The foundation stone that was laid was located at the northeast abutment and into it was placed a glass vase containing a record of the day's events. The scroll read:

> The foundation stone for this bridge was laid on the 10th of November 1859, in the 23rd year of the reign of Her Majesty Queen Victoria, in the year of Masonry 1859, by His Excellency, the Earl of Carlisle. And the Masonic body of the Province of Munster; General Sir George Chatterton, Provincial Grand Master of Munster; George Chatterton Esq. Deputy Provincial Grand Master; William Penrose, Provincial General Secretary.

The vase was placed in a cavity in the huge foundation stone where it remains today.[107] The Lord Lieutenant was then presented with a silver inscribed trowel by John Benson, commemorating his visit to the city for the ceremony. The inscription read:

> Presented to His Excellency the Earl of Carlisle, Lord Lieutenant of Ireland on the occasion of the laying of the foundation stone of St Patrick's Bridge, Cork on the 10th of November 1859; John Arnott, Mayor; Sir John Benson, Engineer; Joseph Hargrave, Contractor; William Barnard, Clerk of Works.[108]

In this inscription lay evidence of yet another piece of historical coincidence; the contractor Joseph Hargrave was none other than the grandson of Michael Hargrave who had played such a large part in the building of the first St Patrick's Bridge.[109]

With the foundation stone in place what lay ahead was the actual construction of the bridge.

Shipwreck

Prior to all of this, another shipwreck had taken place at the site, on Wednesday 21 April 1858. This one, however, did no structural damage to the timber bridge which Benson had constructed. In was a Portuguese ship called the *Funchal* and the master was one Captain Rodriguez. With a cargo of 160 tons of salt for delivery to local merchant John Firmo, the ship was piloted upriver from Queenstown (Cobh today). In those days, however, the river dried out very quickly in the late spring and summer periods. Before the ship had been tied off properly, the tide receded and the ship, which was moored opposite the stores of Messrs Harvey & Co., St Patrick's Quay, listed to starboard until she went over completely and crashed against the quay wall. The entire cargo was destroyed. An attempt was made to right her the following Friday, with the attachment of ropes and 'three or four relays of men from an early hour' trying to pull her upright, the rope snapped and she again crashed down, this time completely wrecked.[110]

Construction Work and another Temporary Bridge

The task facing the builders of the new bridge was a huge one by any standards and the work would cause serious disruption to the traffic and pedestrian flows. In order to minimise this disruption, the Improvement Department ordered that another temporary footbridge be constructed downstream from Benson's Bridge, between Merchant's Quay and St Patrick's Quay, just opposite Hackett's Store. This bridge was described as being, 'ten feet in width and although constructed entirely of timber, is light and ornamental in appearance.'[111] Built by Mr Barnard and contractor Mr Edwards, the bridge took only three weeks to complete and was opened to the public on Friday 27 May 1859.[112] (When St Patrick's Bridge was completed the *Cork Examiner* suggested the timber footbridge could, 'without difficulty be raised and conveyed by

From the Milton Sellwell of London collection. (Courtesy Michael Lenihan.)

barges to the foot of Wyse's Hill where it will be placed and where a footbridge was long been deemed very necessary'.)[113]

With the temporary footbridge completed, the contractors for the main bridge firstly had to remove the foundations of the old bridge.[114] Divers were used in this operation and the blocks of stone which were removed were taken downstream to where the Marina is today to be reshaped for use in the new bridge. Then the foundations for the piers and abutments of the new bridge were sunk into the riverbed to a depth of between 10 and 14ft below the low-water mark.[115]

The next process was the laying down of a layer of cement. This was achieved with the use of specially-designed boxes, fitted with trapdoors that were released when the correct depth was reached. Each layer was 10ft wide and the cast-iron caissons were laid on them. They were then joined together, forming completely enclosed cases around each pier. Further layers of concrete were then laid around these and the first levels of masonry work, which consisted of blocks of stone weighing anything up to 3 tons each, were mounted on top of the caissons.

When the arches and the balustraded top of the bridge were in place, Cork City boasted the widest bridge in these islands with the exception only of Westminster Bridge in London. The total waterway span was 168ft, the centre arch being 60ft with the other two being 54ft each. The width between the parapets was 60ft 6ins, the roadway being 40ft. Limestone was used for most of the construction, the material for the foundation blocks coming from Foynes in County Limerick.

Two serious accidents occurred during the course of construction: a workman fell into the river when crossing between two barges and was drowned and a man fell 30ft from a gantry, sustaining serious but not fatal injuries.

On 12 December 1861, the new St Patrick's Bridge was opened to the public for the first time. Rain lashed down and a southwest gale whipped the river into frenzy. One old woman was said to have remarked that if the Mayor tried to open the bridge he would be swept away. Sir John Arnott, however, serving his third successive year as Mayor, did open the bridge saying, 'I have great pleasure in congratulating the inhabitants of Cork on the completion of this elegant bridge which I now open for public traffic.' Sir John Arnott completed his speech and led the members of the corporation across the bridge, surveying the workmanship. Before they could return, however, a hackney driver drove his nag across, determined to be recorded in history as the first person to cross the bridge.

St Patrick's Bridge is the same today as it was all those years ago, apart from re-strengthening work done in the early 1980s, following which a plaque was placed in the bridge wall containing the names of both a Mayor and a Lord Mayor of the city. The Mayor was Sir John Arnott and the lord Mayor was Councillor Paud Black. (The title of the city's First Citizen became Lord Mayor in April 1900 at the granting of Queen Victoria.)[116] In 1861, the bridge was lit by gas lamps. Today, ornamental lamps that stood high over the bridge on that December day

Effigy on St Patrick's Bridge.

are still to be seen. The effigies on the top of each arch were carved by a sculptor from Douglas Street named Scannell; they are of St Patrick, St Bridget, Neptune and the three sea goddesses.[117]

In summary then, the first St Patrick's Bridge was built late in the eighteenth century, when the city began to expand eastwards. The building of the bridge facilitated both the economic and social dimensions of this development and the elaborate opening ceremony was a stage-managed performance that reinforced the social norms of the time. Seventy years later, the debates and discussions surrounding the building of the second St Patrick's Bridge can be read as a narrative of changing society wherein the public could express their viewpoint and representative groups could argue for outcomes most favourable to their members. Nevertheless, the elite of society again stage-managed the opening ceremony using symbols from the earlier ceremony. The design of the new St Patrick's Bridge and the addition of the ornamental lamps and the sculpted effigies were architectural adornments not alone for the bridge but also for the greater hinterland. Thus, in spanning the River Lee's North Channel, St Patrick's Bridge facilitated social and economic interactivity in the developing city; was an architectural adornment of the city, and finally, was a stage upon which ritual and ceremony was performed for the advancement of the elite in society.

3

1800-1850

> Cork in the 1820s was a city split as much on politico-religious grounds as on those of
> class. It would be too simple to describe the political scene in terms of Catholics versus
> Protestants but there is a rough reality to that picture.[118]

During the previous century, expansion beyond the old medieval town walls had led
to the development of a new centre to the east of the North and South Main Streets
with new bridges leading to developing suburbs to the northeast and the southwest.
This trend continued during the first half of the nineteenth century when a rising
population moved further to the suburbs and yet more bridges were built to enable
these developments. The *Constitution* reported in 1831 that the population of the city
in that year was 106,980, an increase of 6,455 on the figure of 100,535 of a decade
earlier. (An article associated with the figures warned that, 'in every numbering
of the people there is an inherent infirmity which wars against accuracy'.)[119] This
expansion and movement of the population, as well as economic developments, were
primary factors in the bridge-building endeavours in Cork during the first half of the
nineteenth century.

The Second Parliament Bridge

There had been bridges other than those on the river's main channels built in Cork
during the latter half of the eighteenth century: Punche's Bridge over the Kiln River
was one such that stood until near the end of the twentieth century. As 1800 dawned,
however, Cork's main bridges were those over the Lee. On the North Channel, the first
St Patrick's Bridge rivalled the North Gate Bridge as the most important, while on the
South Channel this status belonged to the South Gate Bridge; Clarke's or Wandesford's
Bridge too, was of great importance.

The other South-Channel crossing, Parliament Bridge, was, however, causing
concern. Built in the early 1760s, by 1803 newspapers were stating that aspects of
the bridge's condition were 'truly alarming'.[120] On 28 May 1804, they reported that,
the wooden part of the bridge, 'was this morning carried away by the flood' and the

Parliament Bridge. (Courtesy Michael Lenihan.)

death of a Cork gentleman is recorded for 18 June of the same year, 'he having missed his way among the ruins, fell into the river and was drowned.'[121] A decision was taken to replace the structure and Abraham Hargrave, who had previously been involved with St Patrick's Bridge, was asked to design the new crossing. At a cost of £4,000, the new Parliament Bridge was opened to the public in 1806. This is the bridge still in use today, having been totally refurbished in the 1990s. A single-arch limestone structure spanning 65ft 6ins, the overall width is 40ft. It is a truly hump-backed bridge with beautifully sculptured balustrade on either side through which children love to watch the rushing waters below.

Careful examination, however, reveals that repair work was done at some point in the bridge's history prior to the 1990s refurbishment, when concrete was used instead of cut limestone to repair damaged balustrades. In his study of Cork's South Parish, Roger Herlihy suggests that the bridge may have been damaged in August 1922, when Free State troops sought to take the city from the anti-treaty forces.[122] Reports in the *Irish Times* and the *Freeman's Journal* mention a number of places damaged or destroyed during the fighting, among them Parliament Bridge.[123]

At a meeting of the General Purposes Committee of Cork Corporation, held on Tuesday 4 November 1930, the matter of the condition of Parliament Bridge came up for discussion. The city manager reported to the meeting that a proposal to spend £300 to replace the parapet of the bridge should not be accepted as the bridge was unsuitable for the demands placed upon it; it would need to be replaced within the next five or so years and the money would be better spend on Carroll's Bridge adjacent to the North Channel.

Postcard of Parliament Bridge. (Courtesy Michael Lenihan.)

Parliament Bridge in the twenty-first century.

Councillor O'Sullivan, however, rejected this saying that the repairs to Parliament Bridge had been carried out before the existing corporation had come into office and that there was, 'a general complaint at having concrete put into a limestone structure'. He went on to say that the overwhelming majority of Cork people wanted Parliament Bridge preserved as it would be there,'long after the members of this council had passed away.' Councillor Horgan, agreeing, said that the matter should not even be discussed and Councillor Barry said that he would oppose anything being done to the bridge, which was an old landmark and an ornament and credit to the city. Consequently, the matter was dropped, but the Lord Mayor stated that it could be brought up in the future through a proper motion should parties so wish.[124] The bridge did indeed outlast the members present on the occasion of that meeting and in the early 1990s was revamped such that it was ready to face into a third century.

The George IV Bridge

Post-1800, after the Act of Union, were difficult times throughout Ireland. Catholic Emancipation, which had been much anticipated following the enactment of the legislative union of the kingdoms, was not forthcoming, not least because of King George III doggedly upholding the Protestant constitution that he believed in and refusing any leeway towards Catholic liberty. There was, however, during these first decades of the nineteenth century, an increasing emergence of peasant, and indeed Catholic, awareness of their role in the workings of society, a social awakening that has been chronicled by the historian Bartlett and regarding Cork specifically, Maura Cronin.[125] The emergence of Daniel O'Connell as leader of Catholic Ireland would not bear fruit for some years to come, but by the mid-1820s his movement, the Catholic Association, was already well underway. Consequently, it was with a degree of optimism that the country greeted the coronation of a new monarch, George IV, on Thursday 19 July 1821:

> On Thursday, being the day appointed for the Coronation of His Most Gracious Majesty, George IV, the troops of this garrison fired a Feu de Jove. The same evening the towns of Cashel, Thurles and Cahir were brilliantly illuminated in compliment to the august ceremony.[126]

This was written in the *Freeman's Journal*, which also reported that, 'the Coronation was celebrated by public rejoicings at Balbriggan and Baltinglass,' while, 'a splendid ball was given at Kilkenny'. At Cork there were grand illuminations and a grand dinner took place at the commercial buildings in that city.[127] The illuminations and dinner, however, were not all that occurred in Cork in celebration of the coronation of George IV:

> Fifteen Masonic Lodges, with the master and wardens of the Coopers' trade were in the procession with the Mayor and Corporation at Cork on the Coronation day. The first stone of a bridge, which is to be called 'The Bridge of George the Fourth,' was laid at the south branch of the Lee.[128]

Yet again, the Masonic Lodges were associated with a ceremonial bridge-inaugurating event in the city. That it was specifically a part of the celebration of a royal coronation gave it added significance. Henceforth this bridge, forever associated with his coronation day, would stand as a concrete statement of the loyalty of his subjects: the people of Cork. As with the other bridges in the city, there were also practical considerations in its development.

With the continued expansion of the city, the need for a major route westwards became apparent. This was not only required to facilitate the expanding suburbs, but also the trade coming into and leaving the city from the western reaches of the county and beyond. To this end, the building of what is now known as the Western Road began in the year 1820. This route would always have to cross the South Channel of the Lee, downstream from the point where the river divided to the west of the city. Consequently, it was here that the George IV Bridge was built. The laying of the foundation stone occurred to coincide with the King's coronation. The bridge is a

O'Neill Crowley Bridge, originally George IV Bridge.

three-arch structure, just over 51ft wide and was built by the Paine brothers, George Richard and James, architects who also designed the St Patrick's and Holy Trinity churches.

With the advent of the twentieth century, the desire for independence from Britain was growing throughout most of Ireland. Following the restructuring of local government in 1898, in subsequent elections many councils came to have nationalist majorities. They were, consequently, in a much better position to implement a policy of changing the names of various streets and bridges throughout the country with the dual purpose of seeking to destroy any manifestation of British imperial identity and advancing an Irish nationalist one. In 1907, the corporation in Cork decided to change the name of the George IV Bridge to that of O'Neill Crowley, in memory of the Fenian patriot from Ballymacoda. Despite the approval of this decision, the name was not formally altered until 1912.

Peter O'Neill Crowley was born in the East Cork village of Ballymacoda in 1832, the son of a prosperous farmer. Following an unsuccessful engagement at Knockadoon coastal station during the 1867 Fenian Rising, he and a number of colleagues attempted to reach other units in North Cork. However, he was shot dead in an engagement with police and soldiers at Kilclooney Wood near Kildorrrery on 31 March.[129] As well as the bridge, O'Neill Crowley is also commemorated on the National Monument at the southern end of the Grand Parade in the city.

By 1824, the new western route had reached the Grand Parade with the completion of Great George's Street, subsequently renamed Washington Street after the American President George Washington.

Wellington Bridge

Prior to this time, the suburb of Sunday's Well had been developing, and was for many years an area of some affluence. It was a suburb where an emerging middle class lived in, 'three and four bed-roomed terraced houses' and where at one time twenty-one nobility and gentry out of a city-wide listed total of 158 were resident.[130] However, the construction of the new City Gaol there in 1825 both represented and reinforced a decline in the area's fortunes.[131] The nearest river crossing for the residents of this area to gain access to the city was the North Gate Bridge and this was a cause of concern among those living there. As early as 1804, it was announced in the *Cork Mercantile Chronicle* that, 'a new bridge is intended to be made across the River Lee at the western end of the North Mall'. A reward of five guineas was offered for the best submitted design.[132] Two months later the same paper editorialized that:

> We are glad to be informed that a handsome and convenient communication between the county and the city is about to be made ... The new bridge about to be thrown across from the North Abbey to the corner of Grenville Place, and the improved road from thence into the country through the beautiful suburb of Sunday's Well will prove highly remedial of the inconveniences long laboured under in that quarter of the city... The new bridge is one of the most popular projects we have for a long time witnessed ...[133]

A quarter of a century later, there was still no bridge. However, at the other end of the suburb to the west, a bridge was planned that would connect with the new Western Road out from the city. Today that bridge is named Thomas Davis Bridge. There were objections to the proposal to build this bridge, as part of a beautiful amenity known as the Mardyke Walk would be lost during the course of the construction. This Mardyke Walk had been built in 1719, largely through the undertakings of Mr Edmund Webber, a Dutchman who was at one time High Sheriff of Cork. (He was made a freeman of the city in 1728.) The walk travelled westwards from the city along the banks of the river and went as far as a house with 'good gardens planted with fruit for the accommodation and entertainment of persons who frequented the Walk',[134] situated where the Sacred Heartchurch and residence is today.

Despite the objection, the bridge was completed in 1830. Like the George IV Bridge of 1820, the Paine brothers were taken on to build it. Richard Griffith, a geologist and engineer, better known for his work on roads throughout the country and for land valuations upon which taxation was based, designed the bridge. It consisted of

Wellington Bridge. (Courtesy Michael Lenihan.)

a centre arch of 50ft and two side arches, each of 45ft. The parapets were solid, with the piers of the arches sunk in caissons.[135]

On completion, the bridge was named Wellington Bridge after the Duke of Wellington. However, just like the George IV Bridge before it, this name was changed, to that of Thomas Davis Bridge. Despite that change, very many people still refer to it by its original name.

Thomas Osborne Davis was born at Mallow, Co. Cork in 1814, the son of a surgeon. In 1839, he joined O'Connell's Repeal Association but grew tired of O'Connell's strategies and founded the Young Ireland movement in 1842 with Charles Gavin Duffy and John Blake Dillon. He died on 16 September 1845 and is remembered not just for his writings in the Young Ireland publication *The Nation*, but also for such enduring songs as 'A Nation Once Again' and 'The West's Asleep'.[136]

The Gaol Bridge

The next bridge-building project undertaken in the city was that of the Gaol Bridge, which was completed in 1835. The purpose of this bridge was to give access to the county gaol on the South Channel of the river from the recently-completed Western Road.[137] Up until then, access would have been via the routes extending from Barrack Street and the road to Bandon down through College Road, then known as the Gaol

Road. Lewis wrote that it was a handsome bridge, midway between George IV Bridge and the Lee Mills, which, by a raised causeway, leads from the new Western Road to the county goal and house of correction.[138]

It was towards the end of a six-week visit to Ireland, in the autumn of 1833, that the famous engineer Marc Isambard Brunel visited Cork. According to his biographer he, 'was fêted wherever he appeared'.[139] The visit to Cork was a short one: on his return to London he wrote to a friend Mr Coxson that 'I have, in the course of one week seen Killarney, Cork, Bristol, Bath, London, Ramsgate and London again.'[140] Despite the brevity of the visit to the city by the Lee, however, it was sufficient to leave a lasting mark. While in Cork, he was invited, 'to submit a design for the delightful little bridge over an arm of the River Lee known as Cork Jail Bridge. Richard Beamish had recently been appointed engineer to County Cork and it was through his introduction that the authorities engaged Brunel for the work'.[141]

The bridge is one of a single arch spanning 50ft, has solid parapets and was again made from hewn limestone. It has a width of 31ft 6ins with a single 3ft pathway. The portico at the front of the goal was also built at this time, having been designed by the Paine brothers in 1818.[142]

Between 1800 and 1835 therefore, the corporation was responsible for overseeing the building of three completely new bridges across the Lee, as well as the replacement of another. All of the projects aided, in one way or another, the growth of the city, both in terms of population and spatial expansion. None, though, could be said to have

Gaol Bridge designed by Brunnell.

been undertaken for purely economic motives. However, during this period also, the first Anglesea Bridge was built across the South Channel and this project was certainly economically motivated. Furthermore, it was not entirely a corporation initiative.

Anglesea Bridge

Throughout the 100 years prior to 1830, Cork's trade in grain and corn had expanded enormously. In 1771, the Cork City Corn Market was extended from two to four days per week.[143] However, efforts to manipulate the market by farmers on the one hand, and merchants on the other, coupled with an economic downturn in the post-Napoleonic War years, meant that the industry as a whole was undermined. Dickson describes how a 'motly group of city buyers and their manner of direct or commission dealing ... subverted the principal urban markets ...'[144] Prices collapsed and only then were there efforts to re-establish public markets for grain and corn. New regulations were imposed in the 1820s, following which, in a single year, 75,674 barrels of wheat were exported, an all-time record.

With a new order established, it was decided to build a new Corn Market building. Thus, at a cost of £17,000, the Corn Exchange at Albert Quay was opened in 1833.[145] The new building was located at Sleigh's Marsh on the southern side of the river where the City Hall stands today and it subsequently served as the City Hall until that building's destruction on the night the 'Burning of Cork' on 11 December 1920.

The main problem that the new Corn Exchange presented was its inaccessibility from the city centre. To reach it, it was necessary to cross the river at Parliament Bridge and then go down the quays. Thus, the Corn Market Trustees, a body established by Act of Parliament to oversee the industry, considered the idea of constructing a bridge from Warren's Place on Lapp's Island (now Parnell Place), to the new exchange at Sleigh's Marsh.

The trustees persuaded the owners of Sleigh's Marsh to contribute the land at no cost and to further contribute £2,500 because the proposed bridge would increase the value of their extensive remaining properties. The government was to give a grant of £5,000 and the corporation was also to donate funds. The entire project would cost £9,000 and a well-known engineer of the day, Alexander Nimmo, along with a Board of Trade official, Sir Richard Griffin, were called upon to oversee the operation. After consideration of a number of plans, the one finally decided upon was that of a solid-stone structure with a centre lifting arch and, tenders for the contract having been considered, it was given to Cork engineer Thomas Deane to bring to fruition.[146]

Mayor of Cork in 1830 and knighted in the same year, Sir Thomas Deane was one of a family of well-known architects and engineers living in Cork at the time. He was father of Sir Thomas N. Deane and grandfather of Sir Thomas M. Deane each of whom in turn contributed to the architectural fabric of the city. Among the original Sir Thomas Deane's achievements in the city, were the beautiful buildings for Queen's College (now UCC) and the Commercial Buildings, now part of the Imperial Hotel.[147]

The keystone of the new bridge was put in place at nine in the morning on Thursday 3 June 1830, a fact reported in the *Constitution* newspaper on Saturday 5 June, along with a denial that the bridge was to be called Waterloo Bridge:

> A paragraph having appeared in an Evening Paper stating that the bridge at Lapp's Island, now nearly completed by Messrs. Thomas Deane & Co. was named Waterloo Bridge, we are requested to state that no such occurrence took place by any authorized person. We presume that the bridge when finished will be named by the Trustees of the Market.[148]

It was subsequently named Anglesea Bridge in honour of the Marquis of Anglesea whose reign as viceroy was regarded with approval by many among the people of Cork.

Like others in the city, the bridge was of hewn limestone and had pilasters of cast-iron. The two elliptical arches spanned 44ft, with a rise of 11ft and a waterway of 32ft was crossed by a drawbridge, which would admit shipping.

The first Anglesea Bridge was clearly more than just another crossing point on the South Channel of the river. Its purpose was primarily to make for easier access to the newly developed Corn Market at Sleigh's Marsh and thereby aid economic improvement in the expanding Cork of that time, whilst its design meant shipping could still pass upstream, thereby ensuring that the requirements of the port

Anglesea Bridge in the 1860s.

Anglesea Bridge. (Courtesy Michael Lenihan.)

were still catered for. As seen previously, the chosen name for the new bridge both acknowledged and perpetuated the social order of the day.

The bridge was to remain in service for approximately half a century during which time it was both loved and hated by the people of Cork. The antiquated drawbridge system was a cause of great frustration to shipping-masters, the merchants and the ordinary members of the public, all feeling that they should have priority in its use. It was inevitable, therefore, that during the latter half of the nineteenth century, moves would be made to replace the first Anglesea Bridge and when that happened the name chosen for the replacement bridge again reflected changing times.

New Wooden Bridge

The last bridge-building project undertaken in the first half of the nineteenth century occurred at Castlewhite, not far to the west of the County Goal where Brunnel had built the Gaol Bridge some twenty years earlier. The *Cork Examiner* described how this new bridge provided a handsome approach to the estate there and that the bridge reflected great credit upon the builder, Mr John Barry:

> It consists of an arch of 90 feet span by 12 in width on the self-supporting principle, having a clear sweep from end to end without props or buttresses in the bed of the river. It is light and graceful in appearance, while it is durable in construction, the timber work being strongly knit and secured by iron bolts and rivets and is fully capable of bearing a weight of 10 tons.

The bridge was secured to cut limestone walls on either side of the river, and prior to its opening, was subjected to a strength test, with 5 tons being conveyed across it. With the opening of Lapp's Asylum to the south in 1856 the bridge facilitated access to this home for impoverished elderly ladies. Today a more modern structure provides access from the Castle White Apartment Complex to the University campus.

4

1850-1900

For the first time in Irish history the masses came on the political stage as leading players rather than extras.[149]

Anglesea Bridge: Call for Replacement

Three different bodies were involved in discussions pertaining to Anglesea Bridge during the middle decades of the nineteenth century: the corporation, the Corn Market Trustees, and the Harbour Commissioners. The next phase in the saga occurred in 1863, when Mayor John Francis Maguire called for the replacement of the old Anglesea Bridge at a Harbour Commissioners meeting held in December. There had always been problems with shipping passing through the centre section of the bridge and things came to a head when, on Saturday 10 October, a ship had become stuck in the narrow passageway, holding up all domestic and commercial activity for almost an entire day.[150]

At the Commissioners meeting the following Wednesday, the Mayor asked the harbour master whether, 'the ship which came into collision with the bridge did not strike it on the opening owing to their being want of room for her to pass through?' In reply, Captain Clarke said that, 'yes, that was the cause of the collision certainly.'[151] On 2 December, at another meeting of the Harbour Board, the Mayor moved for a committee to be established to obtain plans and specifications, or plans and information, as to the best manner of rebuilding Anglesea Bridge. He went on to say that a swivel bridge would be desirable and that, although not mandated to say so, he was certain that the Corn Market Trustees would be in accord with his suggestions.[152]

However, when the Corn Market Trustees met the following day, under questioning, the Mayor articulated a different view. He maintained that while the Trustees and the Harbour Board could be of one mind regarding the bridge replacement, they ought not to think of demanding from the corporation anything more than a nominal sum towards the costs involved. When pressed by Mr Bruce as to whether he was saying that the corporation didn't wish to contribute towards the building of the bridge, the Mayor replied that, 'what I say is – I really dare not make a large demand from the corporation.' At the end of the meeting the decision as to whether or not the Trustees were prepared to be involved in the proposal was referred to the finance committee to report on.[153]

Following discussions between the concerned parties, it was agreed that the corporation would match a £5,000 contribution by the Corn Market Trustees towards the cost of a new bridge. When the Trustees realised, however, that there was a strong movement that the suggested bridge was to be a swivel one, they withdrew their offer on the grounds that they would not be involved with a project that would take up large amounts of quayside space. Even after Sir John Benson had produced plans to show them otherwise and the same plans had been approved by the Harbour Commissioners and the corporation, they remained adamant in their refusal. Consequently, no progress was made.

Six years later, in November 1869, the issue was again on the agendas of the associated bodies. At a meeting of the Corn Market Trustees, the financing of the existing bridge was discussed. A report of the finance committee regarding the maintenance of Anglesea Bridge was read to the meeting and included the view that the bridge was now, 'the main thoroughfare into the city from the terminus of five railways, from the Navigation Wall, the Park, the Gas Works, Albert and Union Quays', and that the bridge had encouraged the development of many areas beyond the city, 'thereby largely increasing the area of city taxation'. As a consequence, 'the duty of maintenance lies not with the Trustees but on the corporation'. It appeared that the Trustees would continue with their policy of non-involvement in Anglesea Bridge matters. The report, however, did advise caution, recommending that the Trustees should have a say in the future working and management of the bridge. To that end, a committee was appointed to meet with members of the Harbour Board regarding the future of the bridge.[154]

At a corporation meeting two weeks later it was mentioned that the Trustees and the Harbour Board committees were to meet about the bridge and Mr Keller asked whether the corporation should also adopt the same course to protect the interests of the ratepayers. Accordingly, the Mayor and Messrs Gould, Keller, Nagle and O'Sullivan were appointed to a bridge committee.[155]

Little happened over the following five years; however, in 1874, following the appointment of a joint committee comprising members of the corporation, the Harbour Commissioners and the Corn Market Trustees, a resolution was passed on 5 May that the City Surveyor, Robert Walker and Harbour Engineer Philip Barry, study and report on the various routes and roadways associated with crossing the South Channel of the river. Prior to the publication of the report, on 5 October, the Mayor announced to the corporation that the joint committee recommended a swivel bridge in place of Anglesea Bridge at a cost of £15,000 and an iron bridge similar to the North Gate Bridge, from Grand Parade to Sullivan's Quay. (This iron bridge at the North Gate replaced the stone bridge of 1713 in 1864, as we shall see later in this chapter.) On hearing this, one of the corporation members, Alderman Kelleher, to sounds of great laughter, interjected that, 'this would be the greatest waste of public money ever attempted.'[156]

In fact, the committee's report, as it appeared in the following Friday's *Cork Examiner*, contained four recommendations: firstly, the replacement of Parliament Bridge; secondly, the replacement of Anglesea Bridge; thirdly, a new bridge from the Grand Parade to Sullivan's Quay and finally, a new swing bridge from Charlotte's Quay to Copley Street.[157] The reasons for these proposals were many and varied, among them that Parliament Bridge was unsatisfactory and incapable of meeting the needs placed upon it, partly because of the inadequacies of Anglesea Bridge. They found Anglesea Bridge to have satisfactory ironwork but unsatisfactory stone arches, especially the arch on the south side which suffered from 'serious symptoms'. As a consequence, the report found that 'the safety of the bridge will be seriously imperilled.'[158]

John Francis Maguire, who had started the whole replacement issue in 1863, had died on 1 November 1872,[159] but Mayor Daniel A. Nagle took up the cause. He argued that there were now three railways operating at that crossing, the Cork, Blackrock and Passage; Cork, Bandon and South Coast; and Cork and Macroom Direct Railway; and that between them they carried nearly three quarters of a million passengers a year. For this reason the bridge needed replacing.

His persuasiveness eventually won the day and agreement was reached; the Harbour Commissioners would contribute £10,000 towards the project. The corporation's contribution would be met by a halfpenny increase in the rates and it would promote the necessary legislation in Parliament. Corporation meetings in the first weeks of 1875 were dominated by debates concerning the various measures that should be included in the Bill. Among the more significant elements were the amounts of monies required for the project and the timeframe within which the entire operation would continue.

There was great surprise when, at the meeting held on Monday 18 January, in answer to questions, the Mayor said that £25,000 was to be borrowed of which £5,000 was for a new St Vincent's Bridge, £15,000 for a new Anglesea Bridge, £3,000 for a temporary bridge while Anglesea was being constructed and £2,000 for the expenses of the bill. Mr Fox and others, however, said that this was the first they had heard of a new St Vincent's Bridge.[160]

This matter was again raised at the meeting of 1 February when Mr Cantillon wanted to know whether the sanction of the corporation had been obtained for the expenditure of the extra £5,000 on St Vincent's Bridge. In reply, the Mayor stated that he was not aware of any specific resolutions regarding this but that a petition had been adopted in favour of the bill which embraced both bridges. As the discussion was now drifting away from the Anglesea Bridge issue, the town clerk intervened and explained that two years previously the corporation had agreed to accept contracts for erecting a new iron bridge on the site of St Vincent's, though doing so was in fact illegal and that this now was a mechanism to rectify that situation.

By late spring, the proposed bill was taken to Parliament and the 'Cork Improvement Act of 1875' became law on 2 August in that year.[161] It gave the corporation permission to borrow £25,000 for the purpose of building a new £20,000 bridge, that name of which would be decided at a future date, as well as a new St Vincent's Bridge. There was an important stipulation in the bill saying that that, should the project not be completed within six years of the bill becoming law, the powers granted therein would be withdrawn. The bridge therefore had to be completed by 1 August 1881.

Problems

It was now that the foundations for future problems were laid. Following discussions between the corporation and the Harbour Commissioners, a bridge committee was formed, its purpose being to oversee the entire project. In July 1876, after much argument and discussion, the committee decided to offer a prize of 100 guineas to the best-submitted design for the new bridge. At the meeting, Mr Banks asked whether the corporation was 'bound to go on with the bridge', to which the Mayor replied that, 'it would be very inconsistent not to go on after you have got the Act of Parliament and power to borrow money.'[162]

The advertisement inviting engineers to enter the competition appeared in the newspapers on 8 July. It confirmed the offer of, 'a premium of 100 Guineas for the most approved design', and listed among the requirements that the bridge be capable of carrying, 'heavy road traffic and to be of such strength as to permit with safety the passage of a 35-ton locomotive engine across it.'[163]

The response to the competition came largely from the United Kingdom and no engineer from Southern Ireland entered. It took the committee until July 1877 to decide upon the winning entry. At a corporation meeting on 2 July, the recommendation was made that the prize of 100 guineas be awarded to the plan marked 'A.B.C. No. 9' for the new Anglesea Bridge. Mr Banks, clearly unchanged in his opinion regarding the bridge in the intervening period, interjected to say that, 'the time was come when they should stop the building of the bridge'. However, the town clerk said that was out of the question, following which Alderman Galvin enquired the name of the successful competitor.[164] The prize was awarded to a London engineer Thomas Claxton Fiddler.[165] Progress had been made, but at a price of two of the six years allowed.

Now the committee members acted as if the bridge was going to build itself. They sat back and did nothing for almost another year, eventually submitting the winning design to a Dublin engineer B.B. Stoney. He wrote back indicating his approval of the design and they sat back again, pleased with themselves. Yet more time had passed.

In March 1879, Claxton Fiddler was appointed to three functions regarding the project. Firstly, he was appointed as quantity surveyor for the building of the new bridge so as to enable contractors to tender for the job; he was also appointed the engineer in charge of the new Anglesea Bridge project and finally he was asked to furnish plans for the construction of a temporary structure while the main bridge operation proceeded. The time remaining for the completion of the entire task was now just a little over two years. Alderman Finn welcomed these decisions, pointing out that a recent report had shown that the abutments of the existing bridge 'had gone down six or seven inches' and reminded the corporation that in the not too distant past some people had been killed off St Patrick's Bridge.[166]

The first step was the building of the temporary bridge and in May 1879, Claxton Fiddler produced his plans for this. It was to be a timber construction with a wrought-iron lifting span and machinery for opening same: five tenders were submitted for the contract.[167] These were considered at a Bridge Committee meeting held on 6 June and although the members had been mandated to finalise the arrangements, they chose to bring their recommendations to the main corporation on the following day. At this meeting, Mr Meade sought to have the discussion postponed, stating that he, 'had reasons for doing so that I cannot mention now'. When Mr Sheehan supported this, stating that 'he heard there were some irregularities in the tenders,' the Mayor agreed to hold a special meeting on the following Monday.[168]

At this meeting, the minutes of the Bridge Committee showed that their recommendation was that William Gradwell of Barrow-on-Furness, with a tender price of £2,630 11s 6d, be given the contract. However, the committee had received a letter from Mr Rooney of Queenstown, protesting that his submission had not been looked at fairly. Rooney stated that had he known that pitch pine was acceptable for use in the bridge instead of red pine, as other applicants appeared to have known, he would have entered a lower tender. Claxton Fiddler informed the committee that he had in fact advised three of those tendering that pitch pine was acceptable and would have also advised Rooney of this had he asked the question. At this point, a motion was put to the meeting by Bridge Committee member Alderman Burrows, that while he regretted that a local man could not be given the contract, the price difference was too great and that the committee's recommendation of giving the contract to Gradwell be adhered to.[169] Alderman Dwyer, however, wanted to know why a local man could not be used and sought to have a new tendering process undertaken, based on pitch pine usage. Mr Creedon, in supporting Alderman Dwyer, said that this was a mess that their English engineer had led them into.

This was a clear example of an emerging nationalist identity being reflected in the chambers of the Cork Corporation as evidenced by the tone of the discussion as it continued. At one point Mr Tracy said that 'it was a case of No Irish Need Apply' to which Mr Walker, representing Claxton Fiddler, said that the members were under a misapprehension. Mr Dunlea formally moved that it would be better to advertise again. This motion was carried by seventeen votes to seven, following which Mr Fox and Alderman Burrows resigned from the Bridge Committee.[170]

Just over a week later, at 11 a.m. on Friday 20 June, a joint committee of the corporation and the Harbour Board met to consider the fresh tenders for the temporary bridge. Again there were five applicants: Gradwell's proposal was the same as previously but this time Rooney's was £2,847 10s, £85 cheaper than Gradwells. These were the lowest two tenders and both were formally proposed and seconded.[171] On a division Mr Gradwell's was accepted by six votes to five and the business of the committee concluded.

His celebrations, however, were to be short-lived. Immediately after the joint committee meeting, the corporation met and following the reading of the minutes of the joint committee, Alderman Paul moved their adoption. Alderman Galvin, however, said there was no reason not to accept the lowest tender, that of Rooney. Galvin's proposal that the Queenstown man be given the contract was seconded by Alderman Dwyer, who said that were Rooney's tender £85 more than Gradwell's rather than less, there wouldn't even be a discussion. Mr Tracy supported this view, saying that, 'the idea of giving the contract to a stranger for a higher sum than what it would be done for by a local contractor' was ridiculous. The Mayor stated that, 'he had not the slightest hesitation in such matters in preferring a local contractor to a contractor from England or elsewhere,' justifying this viewpoint on the basis of the local economic situation. Local contractors had not as much work as they ought to have and giving the contract to a local man would help to circulate money among the traders and shopkeepers of the city. The meeting concluded with a vote being taken between the applications of Gradwell and Rooney: sixteen voted for Rooney to be given the contract while ten voted for Gradwell. Mr Rooney was accordingly declared contractor.[172]

The following Thursday, the matter came before the Harbour Commissioners. One final attempt was made to change the situation, with Mr Sugrue proposing that the original decision of the joint committee be adhered to. The motion, however, was defeated by eleven votes to eight. The issue was over; Rooney had the contract.[173] Over the following weeks the formalities were finalised. Rooney named Sir George Miloro and John Steel & Sons as his sureties.

By September, construction of the temporary bridge was underway; at a meeting of the Improvement Committee held on 12 September, Alderman Galvin suggested that the bridge should be called the St Finbarr's Temporary Bridge.

Attention now turned to the bigger job of the permanent replacement for Anglesea Bridge. In October, Claxton Fiddler arrived in Cork. His purpose was twofold: to advise the corporation on the tenders received for the building of the new Anglesea Bridge and to oversee the testing of the completed temporary structure. The test consisted of subjecting the bridge to a weight of 10 tons, under the watchful eyes of Claxton Fiddler, his agent Walker and the City Engineer, Mr O'Keefe. The bridge, 'was found thoroughly up to standard' and praise was heaped upon Rooney and the provider of the iron work, John Steel & Son.[174] Regarding the permanent structure, seven tenders were received for the contract and Rooney was again successful. His price for the job was £16,350 and he immediately started into the project.

Fate, however, was to intervene and prevent him from completing the task. A company in England, the Stockton Forge Company, had been contracted to provide the ironwork, which formed a large part of the new structure. They failed to deliver their materials by their deadline and eventually the problem became so serious for Rooney that on 18 September 1880, he reported to the corporation that the bridge would not now be completed by the due date, 1 August the following year. On 31 December, the corporation informed Rooney that the Anglesea Bridge contract was being withdrawn from him.

Rooney would not accept this. It was the Stockton Forge Company that was at fault and not him, he argued, and refused to give up the contract. The only option available to the corporation was to take Rooney to court. It was February 1881 before an injunction was granted preventing Rooney from proceeding with the job. The case was reported in detail in the *Irish Builder*:

> *The Cork Corporation v. Rooney.* — This case came before the court on motion for an injunction that Alexander Rooney should be restrained from interfering with the engineer and workmen of the corporation, and preventing them taking up the works at Anglesea Bridge, in the City of Cork, for completion. It was also prayed that he should be restrained from withholding certain plant and material used in the construction of the bridge. We may state that Mr Rooney was a contractor for the works, pursuant to the conditions of a contract entered into in November, 1879. The corporation had been authorised by Parliament to reconstruct St Vincent's Bridge and Anglesea Bridge. The contract for the reconstruction of the latter bridge and removing the old one, entered into by defendant, was for a sum of £16,300, the work to be executed to the satisfaction of Mr. Fiddler, engineer to the corporation. In the event of the contract not being fulfilled by the 1st June, 1880, a penalty of £25 a week was to be the result. The contract also provided that if, at any time, in the opinion of the engineer, the contractor had not made sufficient progress with the work, the corporation should have power to take up the work, plant, and material. The corporation engineer did not think that sufficient progress had been made, and a dispute arose, the contractor alleging that he had been entrapped into signing a contract which, in point of fact, was different from the specification.
>
> The court granted the injunction.[175]

Yet more time was lost before the new contractor was appointed. He was the Cork builder John Delaney, a fellow of the Royal Institute of Architects in Ireland and chairman of the Irish District of Municipal Engineers. He promised the bridge committee that he would work around the clock (using the new electric light by night), in order to get the bridge finished on time. His price for completing the job was £13,500 and after a special meeting had been held, he was awarded the contract.

The Bridge Completed

Despite keeping his word that he would work around the clock to finish the job, it was soon obvious that even this would not suffice. By now, eighteen years after the first calling for the replacement of the bridge, the city officials were finally doing something positive. They were pressing Parliament to grant them an extension to the time allowed in the 1875 Act, claiming that extenuating circumstances had caused the problems. On 11 August 1881, the Cork Improvement Extension of Time Act was given royal approval and the following day a letter to this effect was read to the corporation by the town clerk.[176] Delaney was now having the same problems that Rooney had, delays in the delivery of the iron-work and changed specifications.[177]

The project was eventually completed in the second half of 1882 and on 18 November, the bridge was officially opened. Before this, however, a matter of great debate arose, not just in the corporation chamber, but throughout the city that of what name the new bridge should have.

This was a period when an enhanced sense of personal, local and national identity was developing throughout Ireland and Cork was no exception. Joep Leersen has described the post-Catholic Emancipation years as a *Sattelzeit* period, when a sense of identity and historical consciousness developed and was expressed in the public sphere and in public space.[178] Examples of this emergence in Cork, were the huge crowds attending the arrival of the remains of the Young Irelander Terence Bellew McManus in 1861; the destruction of the statue of King George II in March 1862; the placing of a statue to the memory of Fr Mathew on St Patrick Street in 1864, an event attended by a reported 100 thousand people; the Manchester Martyr Commemoration on 1 December 1867 and subsequent years; the Tailor Riots in June 1870 and many more. Thus it was no surprise when, at a corporation meeting on Friday 28 July 1882 to discuss the naming of the new bridge, Alderman Nagle stated that he did not, 'speak in any hostility to the English people or the English Government but this I say that I think it would be more self-respecting if we had called our streets, bridges and avenues after people born among us who did good service for us, and with that view I was anxious that the name of the bridge be changed.'[179] Contemporary reports suggest that the topic had been one of considerable interest among the public for some time.[180]

At a Bridge Committee meeting held on Monday 10 July, St Finbarr's Bridge was suggested as an appropriate name for the new structure and, having been accepted as a recommendation, was brought to the corporation meeting of Friday 14 July. Following the reading of the Bridge Committee's minutes, the Mayor asked whether the members were unanimous in adopting the minutes and thereby the recommendation. Mr Ryan interjected, 'not at all; we won't call it St Finbarr's Bridge.' A discussion then followed, during which Alderman Dwyer proposed that the bridge be called after Charles Stewart Parnell. Sir George Penrose suggested that to name the bridge after St Finbarr was continuing a grand tradition, one that had previously seen bridges named after St Patrick and St Vincent. Mr Dale sought details of the proceedings that had taken place at the committee meeting and it was revealed that of seven members present, five had supported the proposal for St Finbarr while the other two voted for Parnell.[181] As the debate continued Mr Lane said that one of the reasons he supported the Parnell suggestion was that, 'the City of Cork groaned under what he might call shame and degradation in connection with the nomenclature of the bridges, squares and streets of the city. If a stranger came and chanced to look up at the names on the sign-boards at the corners of the streets, he could hardly be blamed for thinking that he was in no Irish city but in Hackney or Westminster.' A vote was then taken and the result was fifteen for the Parnell proposal with thirteen for St Finbarr. The Mayor and Mr Roche did not vote.

First Parnell Bridge by Fergus O'Connor of Dublin. (Courtesy Michael Lenihan.)

First Parnell Bridge in the open position. (Courtesy Michael Lenihan.)

The issue then arose, that with two not voting, yet remaining in the chamber, there was not a majority of the thirty present in favour of the Parnell proposal. The Mayor said that this was a point of municipal law that he could not decide upon himself, but would accept the town clerk's ruling that the resolution was not properly carried. Accordingly, after some debate, the matter was adjourned until the next meeting of the corporation.[182]

That meeting took place on Friday, 28 July. By this time word had spread throughout the city that the issue of what name would be chosen for the replacement of the bridge was under consideration. Furthermore, it was known that the popular view that it should be called after Parnell was under threat, a fact evidenced by public interjections at the meeting. The *Cork Examiner* reported that, 'the gallery was accordingly filled with people who very freely manifested their feelings as each member spoke and voted.'

At the outset, the minutes of the Bridge Committee were again read, including its recommendation that the bridge be called St Finbarr's Bridge, which was passed by a vote of five to two. As at the previous meeting, Alderman Dwyer proposed an amendment that it be called Parnell Bridge. Keller stated that the feelings of every hundred people that you would meet were that the bridge should be named after Parnell. A somewhat humorous dimension to the proceedings then ensued. Sir George Penrose asked whether permission had been obtained from Mr Parnell to have his name brought into the matter. This question evoked considerable debate during which a voice from the gallery asked whether St Finbarr had been asked if his name could be used. For the record, Mr Murphy officially asked what had St Finbarr to say to his name being dragged into the proceedings. Mr Tivy said that he would prefer to see the bridge called Victoria Bridge, 'out of respect for their beloved Queen'. Mr Neehan interjected, 'ah what nonsense,' before Tivy continued, denying that it was the feeling of everyone that it should be called Parnell Bridge and also denying that this was the feeling of the trades. In fact, he said, he had a report from a meeting of the Cork Boot Company to that effect, at which trades secretary Michael McCarthy had been present.

It was then brought to the attention of the meeting, that as far back as three years earlier the name of St Finbarr had been discussed. However, the Mayor pointed out that that had been a suggestion for the naming of the temporary bridge and so had no standing in the present discussion.[183] Finally, a vote was taken and the result was twenty for Alderman Dwyer's amendment that it be called Parnell Bridge with seventeen against. Immediately, there were cheers from the gallery. The town clerk, however, said that this amendment merely displaced the Bridge Committee's recommendation and that a further proposal would now be needed to secure a name for the bridge. Mr Sheehan immediately proposed that it be named Hibernian Bridge. This, however, was defeated by twenty-one votes to nine. As the amendment to replace St Finbarr with Parnell had been successful and that to replace Parnell with Hibernian was unsuccessful, the Mayor declared the proposal that the bridge be called Parnell Bridge, adopted, to loud applause.

The story of the naming of the new bridge is significant on many levels. Consideration of those taking differing viewpoints in the corporation chamber demonstrates the political divide between those in support of continued union with Great Britain and those who were a part of the emerging nationalist identity of the time. The content of many of the contributions to the debate revealed that in Cork, as elsewhere in the country, a reclaiming of public space was perceived as an essential part of this emerging nationalist identity.[184] The crowded gallery illustrated the interest of the ordinary man in the street in the issue and the reported reactions to the proceedings clearly showed where their support lay. It was also important to interested groupings that their position not be misrepresented. On Wednesday 2 August, a letter appeared in the *Cork Examiner* from Michael McCarthy, secretary of the United Trades Association, contradicting 'the audacious assertion made by Mr Tivy T.C. at Friday's meeting of the Town Council "that the trades were averse to having the new bridge named after Mr Parnell".'[185]

The Bridge Finally Opened

The *Cork Examiner* of Monday 20 October, carried a full report on the Parnell Bridge proceedings on 18 November 1882, 'After a delay which became simply exasperating to all concerned in the structure, the Parnell Bridge was thrown open for public traffic on Saturday'. Although it had been announced that there would be

Passing trams at Parnell Bridge. (Courtesy Michael Lenihan.)

Postcard First Parnell Bridge.

no formal ceremony other than the attendance of the councillors at the opening, by early afternoon, 'the approaches to the bridge at the north and south sides became thronged with people and two fife and drum bands, SS Peter & Paul's and Cat Lane were present.'

At three o'clock, Mayor Daniel Joseph Galvin, along with some councillors and officials, made their way from the South Mall to the bridge but in the crowd the Mayor became detached from the rest of the 'informal procession'. All resulted 'in a great deal of confusion'. Eventually Mr Walker asked the Mayor if he desired the centre of the bridge to be swung into position. This was duly completed and the Mayor made his way onto the bridge. After traversing the bridge and returning with some difficulty to the centre, he was handed the keys to the lever-house by Mr Walker. The Mayor then addressed the crowd and named the new bridge Parnell Bridge and formally declared it open. His speech reinforced the desire of sections of Cork society at the time to be associated with the dominant movement of Parnell:

> By the voice of the citizens expressed through their representatives in the Council, it has been determined to associate with this new and creditable structure the name of one who as an Irishman of conspicuous ability and a member of the legislature has won the gratitude and the admiration of the great mass of the Irish people and engraved his name indelibly upon the history of our country.

The Mayor went on to say that at the unveiling of the O'Connell monument in Dublin on 15 August, he had the opportunity to invite Mr Parnell to attend the opening of the

A busy quayside and Parnell Bridge early in the twentieth century. (Courtesy Michael Lenihan.)

bridge. However, the delay in bringing the project to completion meant that Parnell could not attend but he did promise that he would visit Cork again in the near future. That, the Mayor explained, was why he considered it appropriate not to have any formal ceremony associated with the opening of the bridge at this time. Then he called for the chains to be removed, so that traffic could use the bridge, following which, a huge crowd rushed across. If there yet remained any doubt as to where the loyalties of the majority of Cork people lay, they were dispelled when there ascended from the northern suburbs a large balloon in green and white with the motto 'Hurrah for Parnell' emblazoned on it.[186]

Further Difficulties

The following Friday (24 October), a notice, which was dated 17 November, appeared in the local press outlining the regulations pertaining to the use of the bridge, issued by Mr Alexander McCarthy, Town Clerk of Cork. This pointed out that there were two lanes on the bridge and that users should keep as near as possible to the left-hand side; that un-sprung carriages etc. should proceed at as near to walking pace as possible on pain of a forty shilling fine; that for persons riding at the centre of the bridge, the maximum permitted speed was four miles per hour; that the same speed limit applied to sprung carriages etc.; that pedestrians must keep to the right-hand side and walk only on the pathways; and finally that when the bridge was open for shipping, nobody was permitted onto it.[187]

The bridge soon became part of Cork's everyday life. In time, the electric trams which commenced operation in 1898, made their way across the South Channel of the river on Parnell Bridge, just as their predecessors, the horse-drawn trams of the early 1870s, had done on Anglesea Bridge at the same site.

If the corporation thought their troubles were over with the opening of the bridge, they were soon proven wrong. For a considerable period in the lead up to the completion of the bridge, the issues that had led to the dismissal of Alexander Rooney from the contract had been before the corporation.[188] As well as this, Rooney pursued a policy of outlining his grievances in public, in the form of a series of letters to the local press. In one such letter, dated 18 September, Rooney said that, 'all my money (over £3,000), was invested in the contract works of Anglesea Bridge and by the damage done to my position by the corporation retaining the money due to me, I become a bankrupt ...' He then accused the town clerk of 'writing threatening letters to my creditors ...'[189] Rooney's letter was written with an angry tone and another from him, published on 30 September, was even more so. Referring to the division of traffic coming onto the bridge and an opinion by Claxton Fiddler that a lamp-post would do the job of two policemen, Rooney stated that, 'to place a large lamp-post in the centre of the entrance to a narrow bridge with heavy traffic coming up a steep incline and having to turn at right angles over the bridge with only a span of 10ft between the base of the lamp-post and wheel

Base of Parnell Bridge lamp standard.

guard; if all his other opinions on the bridge question are in keeping with this one, I shall leave the public to judge.'[190]

A corporation suit against Rooney's bondsmen, Sir George Miloro and Mr M.B. Dodds, for compensation against their losses due to Rooney's failure to comply with the bridge-building contract's conditions, was met with a threat from Rooney to sue both the corporation and Claxton Fiddler for slander and libel. After lengthy discourse between the two sides, the result was a settlement between Dodds and the corporation (Miloro having died in the meantime), that £1,300 compensation be paid and Rooney agreed not to proceed with his threatened action.

Still the corporation was not finished with problems regarding Parnell Bridge. A dispute now arose with the Harbour Commissioners over payments towards the cost of the bridge and its maintenance. Eventually, a settlement was reached allowing that, if the corporation thought fit, they would make a contribution towards maintaining the moveable section of the bridge.

With that, the saga that had started in the 1820s came to an end, at least for another eighty years. Then the bridge would be replaced again ... but that's a story for another chapter.

From the earliest proposals to replace the first Anglesea Bridge in the early 1860s to the completion of the first Parnell Bridge in 1882, the story weaves its way through a highly significant period in the nation's history. These were the years of Fenianism

Lamp Standard from the first Parnell Bridge.

Parnell Bridge and some of the spires of Cork. (Courtesy Michael Lenihan.)

Happy Christmas from the North Gate Bridge. (Courtesy Michael Lenihan.)

as well as the emergence of Charles Stewart Parnell as a force in Irish and English politics and the fact that the Parnell faction was victorious in the naming of the bridge associates Cork with the ever-increasing Parnellism of the day.

The North Gate Bridge Again

We have already seen the St Patrick's Bridge story and how the destruction of the 1789 bridge in 1853 had been largely blamed on structural aspects of the North Gate Bridge. For this reason, it had been decided when applying for Parliamentary permission to replace St Patrick's Bridge to seek the replacement of the North Gate Bridge as well and the 1856 Act (the Cork Bridge and Waterworks Act) included provision for this project.

At a select committee meeting in the House of Commons on Tuesday 6 May 1856, it was stated by Sir John Benson that, 'the present construction of the North Gate Bridge was an inconvenience to the traffic across it and a considerable obstruction to the flow of water. The river above the bridge was ninety-seven feet wide and widened below to one hundred and ten feet.' Referring back to the flood of 1853 he stated that, 'on the occasion of the great flood ... the water above the bridge was five feet and six inches higher than that below ...'[191]

The structure that had contributed to the problems on that fateful day in 1853 was the stone one built by Coltsman in 1712. It had been repaired and widened in 1831 but otherwise remained the same. While the St Patrick's Bridge project had been underway, although there was less emphasis on the building of the North Gate Bridge, there had been some consideration of the project. On occasions, petitions had been brought to the corporation by traders in the North Main Street advocating, for example, that a suitable temporary bridge would have to be secured before the old bridge could be taken down.

Suitable or not, during November 1857, there was under construction at the White Street Timber Yard, 'the different portions of a foot passenger bridge' which it was intended would carry the several water and gas pipes that crossed the river, during the rebuilding of the North Gate Bridge. Supervised by John Benson, the bridge consisted of two girders, each 106ft long with a rise of 10ft to the centre. Crowned Memel timer was used and the joints were protected with red lead paint.[192] The width of the bridge was 8ft, with the pipes suspended underneath.

On Saturday 27 February 1858, the bridge was floated into place, 30ft to the west of the North Gate Bridge. The sections of the bridge were placed on a barge which was manoeuvred into position on an ebb tide. When the sections of the bridge were accurately positioned, the barge floated away and the bridge settled into place. A report in the *Cork Examiner* stated that, 'when its services are no longer required there, it will be moved up to the foot of Wyse's Hill, which it will connect with Grenville Place'.[193] Some months before this, John Francis Maguire, at a meeting in December 1857, stated that, 'he considered it of the first importance that after the completion of St Patrick's Bridge, it [the North Gate Bridge] should be completed as soon as possible.'[194]

In August 1861, Councillor Fitzgerald moved that steps be taken for the erection of the North Gate Bridge, but Councillor Sheehan stated that he didn't think one was needed. Councillor Scott reminded the meeting that the time frame dictated by the 1856 Act for the rebuilding of the North Gate Bridge was running out and also that there was an ongoing need for a bridge at the foot of Wyse's Hill and another between the North Gate and St Patrick's Bridges. At the end of the debate Mr Fitzgerald's motion was carried.[195] Three months later, an editorial in the *Cork Examiner* called for the operation to get underway, saying that the rebuilding 'must be considered without delay'.[196]

The North Gate river crossing was still as important to the trades and public of the north side as it had ever been, despite the building of St Patrick's Bridge. With the completion of this bridge in 1861, thoughts did indeed turn again to the replacement of Coltsman's structure of 1712. In March 1862, reports from John Benson and Robert Walker were published. Benson confirmed that, 'the obstructions which this bridge presents to the floods are caused by the smallness of the arches; if rebuilt in one span, the bed of the river can be lowered some 3ft which, with the free run through one arch, will in very great measure relieve the Hammond's Marsh portion of the city from injury from the floods'. Walker reported that he had found, 'a fresh crack in the land-arch, North Side. The centre of the fan arches at north and south ends of the bridge appear to have sunk about three to four inches'.[197] Within a week, the decision was taken to proceed with the project and this led immediately to a familiar debate surfacing one more time; whether the bridge should be of timber, iron or stone.

Looking west over the North Gate Bridge. (Courtesy Michael Lenihan.)

In June, the Mayor told a meeting of the Improvement Committee that if a single-arch stone bridge was to be built, it would require the removal of some housing in order to construct the large buttresses. Councillor Sheehan stated that he hoped the iron for the new bridge would be manufactured in Cork.[198] Three days after this meeting, a memorandum was received by the corporation from the inhabitants of the North Main Street that the bridge would be of stone and another to the same effect came from the Stone-Cutters of Cork.

It was clear that an orchestrated campaign in favour of stone was now underway. On 20 June, a number of letters were published to the effect that the bridge should be of stone. Thomas Jennings wrote that, 'it ought to be a three-arched stone bridge; it ought not to be a single-arched iron bridge.' James O'Regan of Blarney Lane wrote that, 'all those suspicious apprehensions and inconveniences attributed to the present structure ... can be easily remedied by replacing it with stone bridge', while Civis Antiques was trusting 'that our tradesmen will never be wronged or our city disgraced by an iron bridge.'[199] Meanwhile, the Harbour Board decided that they would contribute nothing to the removal of the old bridge unless the new one considerably improved the bed of the river.[200]

The issue continued throughout July. At the first Improvement Committee meeting of the month, Councillor Sheehan read a notice that, 'if there is an iron bridge to be built at the North Gate we will give it to the Cork manufacturers as they give employment to the people of the city.' When the corporation met on 14 July, the members were informed that the Harbour Board would not contribute to the project. John Benson told the meeting that a stone bridge would cost £9,629 in total while an iron bridge would cost £6,100. Mr Carroll then, in the course of a long speech, moved that a three-arch stone bridge be built. During his speech, he referred to aspects of the building of St Patrick's Bridge which John Benson said were incorrect. When Carroll tried to make further contributions to the meeting, an argument ensued. Eventually, a number of the councillors left and the meeting was adjourned.[201]

This prompted Thomas Jennings to again write to the press, saying that, 'the majority of the Town Council were afflicted with iron mania' and suggested that, 'public jobbing is sufficiently bad but this insane, imbecile mode of transacting public business is still more destructive of public finances'. He urged the corporation to look to the experiences of Dublin and the costs there of an iron bridge and finished by stating that, 'a dislike to give local employment and utter carelessness of the pockets of the over-taxed rate-payers and disorderly proceedings mark the conduct of the corporation of Cork.'[202]

On 11 August, a public meeting was held in the County Courthouse attended by a large number of people, which the *Cork Examiner* described as largely artisan and other classes of citizens.

Maurice Murray was elected to the chair and opened proceedings with a lengthy speech, during which he referred to another meeting some years previously when the same issue, that of whether iron or stone should be employed, arose in the rebuilding of St Patrick Bridge. He congratulated those present and said that their purpose was legitimate, as they formed a large body of rate-payers who differed from a small majority of the corporation. They were present in the County Courthouse because of the refusal of the city High Sheriff to grant access to the City Courthouse when asked to by a delegation of rate-payers and this action received much criticism from those in attendance. When Mr Murray was finished, he called on Thomas Jennings to address the assembled crowd. Jennings too, spoke at length, also referring to the St Patrick's Bridge experience and suggesting that experiences in Dublin supported their case for a stone bridge. He also stated that Westminster Bridge, which was an iron one, had cost £140,000 over the estimate. At his request, Mr Atkins read a report which dwelt on the disadvantages of iron bridges, 'in consequence of their tendency to vibration, their expansible and contractible character from the heat and cold and other qualities.' The meeting adopted a number of resolutions in support of stonework for the bridge and finished with the appointment of a committee consisting of Messrs Jennings, Barrett, Scraggs and Allen to express the sentiments of the meeting to the corporation.[203]

Letters continued to appear in the press over the following weeks. Then, at a meeting of the corporation held on 1 September, the high sheriff responded to the accusations made at the public meeting regarding his refusal to give access to the City Court House. There was, he said, no deputation of rate-payers, but one consisting of three stone-cutters. As a member of the committee dealing with the bridge issue and also as an officer of the city, he 'thought it would be a course against the interest of themselves and their city if he gave the Court House to a body of tradesmen for the purpose of intimidating the Council and preventing its free action'. Later in the meeting, those members of the committee selected to bring the outcome of the public meeting to the corporation, attended and outlined the views expressed there. As the meeting had over-run its time, the matter was adjourned until later in the month.[204]

On 22 September, John Benson laid plans before the corporation for consideration. In his opinion, 'one arch spanning the river without projection or piers' was best to meet all the requirements, made of stone and cast iron.[205] Benson was questioned in detail about the plans but no final decision was taken owing to the complexity of the matter. The issue came before the corporation again on 29 September. At this meeting, the minutes of the previous meeting, which included a proposal that Benson's plans for an iron bridge be accepted, were proposed for adoption. Immediately there was opposition to this. There followed a protracted discussion, during which a number of amendments were proposed seeking to force the corporation to opt for a stone bridge. Eventually, when these had all been rejected, the original minutes and proposal for an iron bridge were accepted.

One reason for this acceptance was that John Benson now informed the corporation that 170ft of quayside would have to be rebuilt, as well as stone abutments and this would provide considerable employment for the stonemasons of the city.[206] In a letter to the press on the following Wednesday, Thomas McCarthy, secretary to the Society of Stone-cutters, asked that Mayor John Francis Maguire use his influence to get the work underway as soon as possible.[207] For that to happen, however, certain procedures had to be followed and the plans had to be approved by the Admiralty in London. To get to the Admiralty, they had to be approved by the Harbour Board, whose responsibility it was to transmit them to London. This approval was secured on Friday 3 October, following which John Benson wrote to the Mayor on 6 October confirming that the plans were on the way to London.

Early in the New Year 1863, approval having been received from London, Barry McMullen was appointed contractor for the bridge and was told by the Improvement Committee that he should employ people as soon as possible.[208] Advertisements were placed seeking tenders for the iron work and at a corporation meeting on 27 January, Councillor Hegarty asked why advertisements had appeared in Glasgow papers when it was understood that the iron work would be executed in Cork or Dublin. The Mayor, however, suggested that, at the very least, a quote from Glasgow would serve as a good guide as to what would be a fair charge. Fourteen tenders were received, four of which were from local firms. At a corporation meeting held on 2 March, the recommendation of the Bridge Committee was that the tender of Rankin & Co. of

Postcard of the 1864 North Gate Bridge from the Emerald series of postcards. (Courtesy Michael Lenihan.)

Liverpool, at £3,295, be accepted and after some initial objections that the contract was going to a non-local firm. Following the revelation that even the Cork firms would have had to source their iron in England and that considerable monies would still be spent in Cork anyway the Rankin tender was accepted.[209]

The foundation stone for the new bridge was laid in April, but, as work progressed, some people expressed dissatisfaction with the undertaking.[210] Mr Falvey, a publican on the North Main Street, informed the corporation, 'that in consequence of the works at North Gate Bridge, his house was in danger of falling. Mr Carroll of Bachelor's Quay complained to the Harbour Board, 'of obstructions to his business arising out of the rebuilding of the North Gate Bridge.'[211] Nevertheless, on 23 September, John Benson reported to the Harbour Board that, 'works at the bridge were progressing satisfactorily', giving details of the amount of quayside completed to date.[212]

The new bridge was to consist of a series of cast-iron ribs, with cast-iron plates covered in asphalt and concrete, laid as the roadway.[213] The building operation was overseen by Jerome J. Collins, a man far better known for his exploration of the Arctic regions: an expedition on which he perished and from which his body was brought and buried in Curraghkippane cemetery, just off the Blarney Road.

A number of features were prominent on the completed bridge, among them ornamental railings which curved along either quayside; ornamental lamp-posts in the centre and at the ends of the bridge; and iron medallions of Queen Victoria, Albert the Prince Consort, Daniel O'Connell and Sir Thomas Moore. Those of the Queen and the Prince Consort were on the western side of the bridge while O'Connell and Moore were on the eastern side. The bridge was 106ft in length with an overall width of 40ft which included two 8ft footpaths. The incline was one in twenty, compared with one in seven on the previous bridge.

The entire operation was completed in only eleven months and on St Patrick's Day 1864, the Mayor, John Francis Maguire, with the designer Sir John Benson and Mr Barry McMullen the contractor, walked across the new bridge and declared it open to the public:

> The new Northgate Bridge was formally opened this morning at nine o'clock by the Mayor. The bridge was at once availed of for purposes of traffic. Owing to pressure of our space, we defer any description of the bridge, further than that given in our leading columns, until tomorrow.[214]

The new bridge was a wonderful architectural adornment and the *Cork Examiner* of 18 March described the bridge as, 'handsome, spacious and commodious ... it affords the greatest facility for the traffic over it ... and the freest passage for the river that flows beneath'. The piece reminded the people that iron had been the right choice and that, 'a large amount of employment was given to the stone-cutters on the quays and abutments'. Finally the paper congratulated Mr Barry McMullen, the chief contractor, who gave 'his mind and his heart to the work'.[215] A more detailed description of the bridge appeared in the *Cork Examiner* of the following day, outlining the enormous specifications of the bridge, including 80ft long foundations, 24ft piles 6ft apart, as well as a surface of cast-iron covered with asphalt over which was laid a bed of concrete.[216] Both the reports, however, had one important omission, one which was corrected on 19 March. 'We omitted reference yesterday to the exertions of Mr Collins, Clerk of Works in the building of the new bridge. Mr Collins had to remain up for numerous nights in connection with the work while in other ways his labours were of an extremely arduous description ...'[217]

Just as the Parnell Bridge story would reopen in the second half of the twentieth century, so too would that of the North Gate Bridge.

As we can see, the second half of the nineteenth century had been an extremely busy period for the corporation in their involvement with bridges. They had erected a temporary bridge (Benson's Bridge), a relief footbridge, and the new St Patrick's Bridge all at that site. Another temporary bridge had been built while the construction of the new North Gate Bridge was in progress. Further to this a temporary bridge and the new Parnell Bridge had also been constructed at the Warren's Place (Parnell Place), Sleigh's Marsh river crossing.

St Vincent's Bridge

One group of the city's residents were still not satisfied, however. These were the people of the rapidly expanding suburb of Sunday's Well. When an advertisement appeared in the *Cork Mercantile Chronicle* of Monday 2 January 1804 seeking tenders for a new bridge at the western end of the North Mall, followed, on Friday 9 March 1804, by a report that, 'the bridge about to be thrown across from the North Abbey to the corner of Grenville Place ... will prove highly remedial of the inconveniences long laboured under in that quarter of the city', it was clear that the newspapers believed that the event was certainly going to happen and also, that it would prove to be a highly popular project.[218] The bridge, however, was not built for a further six decades.

During those years, the residents of the Sunday's Well area continued to seek the building of a bridge and finally, during the decade of bridge building that was the 1850s, the corporation seemed once again to be willing to listen to their pleas.

When the temporary wooden footbridge, that would also carry service pipes beneath it, was put in place 30ft to the west of the North Gate Bridge in 1858, it was suggested that after it was no longer needed at that site, it could be moved upstream to a location at the end of Wyse's Hill. A year later, in May 1859, when another temporary wooden footbridge was built between Merchant's Quay and St Patrick's Quay, to give

St Vincent's Bridge from the Cork Pictorial Card Co. (Courtesy Michael Lenihan.)

relief to the overburdened Benson's Bridge, it was also suggested that that bridge could be relocated to the western end of the North Mall after the completion of the new St Patrick's Bridge.[219] There were those who also believed that Benson's Bridge itself should be relocated upriver at this time. At a meeting of the Improvement Committee on 9 August 1861, during a discussion about the North Gate Bridge project, Mr Scott reminded the members of the necessity of a bridge at the foot of Wyse's Hill. Later in the discussion, when the matter of moving Benson's Bridge upstream was mentioned, Mr O'Flyn said that, 'the timbers of the bridge were now rotten', which was responded to with cries of 'no, no'.[220] Thus there were two or three possibilities for consideration in the matter of placing a bridge between the North Mall and Grenville Place at the foot of Wyse's Hill.

Early in 1862, the Improvement Committee was told that the Bridge Committee had requested Sir John Benson to have plans and specifications prepared for removing the footbridge at once to the foot of Wyse's Hill. As a result, on Saturday 1 February 1862, an advertisement appeared in the *Cork Examiner* informing builders that the Improvement Department would, 'receive proposals from persons willing to contract for removing the footbridge from its present position east of St Patrick's Bridge and re-erecting same across the river at the foot of Wise's [*sic*] Hill, communicating with Grenville Quay in accordance with the plans and specifications now lying for inspection at the office of the city engineer, No. 20 South Mall'.[221]

The bridge to be moved could not have been that to the west of the North Gate Bridge, put up in 1858, because the advertisement stated that the bridge in question was that to the east of St Patrick's Bridge. Neither was it Benson's Bridge because this was not just a footbridge and also it was located to the west of the St Patrick's Bridge structure. Furthermore, at corporation meetings held on 1 March and again on 29 March, suggestions were made that Benson's Bridge should be moved to a position opposite Mulgrave Road and these suggestions were only made after the timber footbridge had actually been placed at the foot of Wyse's Hill. Of the three possible bridges that could have been moved to the foot of Wyse's hill, the one in question had to have been the one placed between Merchant's and St Patrick's Quays in May 1859. On 21 February, Mr Sheehan complained at an Improvement meeting of the position in which the footbridge *was to be* placed at the foot of Wyse's hill. Then on 9 March, Mr Hegarty commented on the frailty of the new timber footbridge that *they now had* at Wyse's Hill. This tells us that the bridge was moved to its new home between 21 February and 10 March 1862.

Two weeks later, at a meeting of the corporation, Councillor Hegarty proposed that, 'the new footbridge at the base of Wyse's Hill be called St Vincent's Bridge.' The proposal was immediately agreed to. Calling it St Vincent's would connect the bridge with the recently undertaken construction of St Vincent's church at the top of the hill, a church that had been designed by Sir John Benson.[222] After almost sixty years, the residents of the expanding suburb of Sunday's Well had a bridge that made their passage to the city considerably easier.

St Vincent's Bridge in the twenty-first century.

When, in the early 1870s, it was decided to replace the bridge with a more permanent structure, provision for just such a project was included in the, '1875 Cork Improvement Act' that primarily concerned Anglesea/Parnell Bridge. Despite this, problems arose. As stated earlier, at a corporation meeting in February 1875, Alderman Daly and Councillor Cantillon said that they opposed the allocation of monies for the rebuilding of St Vincent's Bridge and that this was the first time they had heard of the issue. The town clerk responded, however, saying that about two years previously the corporation had agreed to take contracts for the erection of a new iron bridge at the site of St Vincent's, but that had been illegal at the time, in consequence of which nothing had been done.[223]

A week later, the Mayor reported that he had now refreshed himself as to the issue; that on 14 December 1874, a resolution had been adopted, 'that the seal of the corporation be affixed to the petition for leave to introduce a bill to enable the corporation to rebuild Anglesea and St Vincent's Bridges.' Following the Mayor's reporting of this, however, a considerable argument ensued, because on behalf of a number of the councillors, Mr Gould said that, 'on that occasion there was no mention of any bridge but one – Anglesea Bridge'. As the debate continued Alderman Daly stated that, 'there was no doubt that it was a convenience to certain gentlemen residing at Sunday's Well and it was while their attention was concentrated on the question of Anglesea Bridge and the engineer's report thereon, that in committee they slipped in St Vincent's Bridge.' Eventually, the town clerk brought an end to the argument by saying that regardless, the bridge as it currently stood was in fact illegal and that if the issue wasn't resolved, they may have to remove the bridge completely and that would cause considerable problems. This brought matters to a close with an acceptance that the project would now go ahead.

The same delays that plagued the rebuilding of Anglesea Bridge seemed also to affect the St Vincent's Bridge project; it wasn't until the spring of 1877 that further progress was made. In May, a deputation of Sunday's Well residents waited on the corporation, seeking a temporary bridge to be put in place while the rebuilding project went ahead, but they were unsuccessful. Early in June, the Standing Committee recommended to the full corporation that the contract for the bridge be given to Messrs Brettel & Co. from Worcester, that being the lowest tender. The town clerk further stated that the contract required that a certain brand of iron be employed and that Messrs Brettel were the only firm that offered to supply it. The Mayor said that Brettel's were currently building a bridge at Ilen Valley near Skibbereen and that Mr McCarthy Downing was security on that. Therefore, Mr McCarthy Downing, as well as many others, knew the firm to be reputable.

Some corporation members, however, wanted the contract given to local firms and particularly so when the Mayor, in answer to a question by Councillor Gould, stated that the cost of the bridge would be £1,778, of which, according to Mr Hegarty, £1,200 would go on ironwork. Mr Tracy proposed that the matter should be sent

The original timber St Vincent's Bridge, erected in 1862. (Courtesy Michael Lenihan.)

back to committee to ascertain whether a local contractor could be employed. On a division, however, this was defeated by nineteen votes to eight and the minutes, including the contract recommendations, were adopted.[224]

By December, however, dissatisfaction with progress was noted at the corporation. Alderman Hegarty said that, 'the work was not progressing at all; there was nothing doing; a large piece of iron had fallen into a hole in the water eight feet deep and had not yet been got out.'[225] Further complaints were made in February when Alderman Hegarty again said that there were not enough men employed on the work; that the contractor was building bridges elsewhere at present and the men who should be engaged at St Vincent's were engaged there. In response, Mr Walker, the City Surveyor, said that the last of the piles would be driven that day.[226]

Three months later, on Friday 24 May 1878, the *Cork Examiner* reported that the bridge would finally open but, 'as yet the shadow of uncertainty hangs over the hour and even the day when the consummation so devoutly to be wished will be vouchsafed us – as on enquiry yesterday we were informed that the opening ceremony will be gone through on either Saturday or Monday.'[227] In the event, it was opened on Saturday 25 May 1878 at eleven o'clock, when the Mayor, accompanied by his sergeants and his secretary as well as the town clerk and the city surveyor, walked across the bridge. In his speech, the Mayor said it was a matter of no little satisfaction that nobody had been injured during the construction, which reflected great credit on the contractors.[228]

The bridge was 186ft in length and consisted of three spans of double lattice wrought-iron girders. The super-structure was supported by two pairs of steel caissons, each 2ft 4ins in diameter, filled with concrete, while the pedestrian roadway was concrete on top of a steel base. On the sides of the bridge, the lattice work of the transverse girders was 6ft high and 18ins wide and the width of the bridge was 12ft. In the quay wall on the south side of the bridge a plaque says 'St Vincent's Bridge, Built 1875'.

The bridge still stands over 130 years later, despite being threatened by the floodwaters. Between seven and eight o'clock on the evening of Monday 21 November 1892, the flood:

> swept by this [St Vincent's] bridge at an alarming rate – the velocity must have been at least ten miles an hour. The force of the current swept by the southern pier and caused extreme vibration, but it is understood that the stability of the structure is in no way endangered.[229]

In December 2009, sections of the quay walls adjacent to the bridge were destroyed by flooding but the bridge itself, despite the concerns of the authorities, stood firm.

As the city and her people slowly edged their way towards the beginning of another new century, thoughts were turning to the factors that would most effect the development of the city in the coming years. Without doubt, the continued expansion of the railways would be one such factor. Another was the birth and development of the motorcar. Both of these would have their effects on the river bridges of Cork City.

5

THE TWENTIETH CENTURY

... naming and more significantly renaming became an important tool in asserting a sense of Irish national identity.[230]

Donovan's Bridge

It was to better facilitate the railways as well as to provide further down-river crossings that Clontarf and Brian Boru Bridges were built and officially opened on 1 January 1912. However, before the twentieth century was even that old, yet another new bridge was built on the South Channel of the Lee. Donovan's Bridge, was situated where Donovan's Road joins the Western Road at the University Gates.

Donovan's Bridge in the twenty-first century.

Made of limestone with an arch span of 60ft and 29ft wide, the bridge was designed by W.H. Hill & Son, engineers in the city at the time, and built by Patrick Murray of John Street. What at first was thought to be a relatively simple job, proved in fact to be extremely difficult. Larger and stronger abutments than originally planned had to be built and to do this, the riverbed itself had to be excavated. It was also found necessary to drive pile at the northern side of the construction. The sheeting over the arching between the voussoirs[231] was composed of immense red sandstone flags, some weighing up to half a ton; these were brought from the Lower Glanmire Road Quarry.

The project came about because land between the river and the College Road, having been laid out for house-building purposes by Mr Thomas Donovan of Fernhurst Avenue, it was necessary to make a connecting bridge between the College Road, the Western Road and these new housing developments. The venture was described by the *Cork Examiner* as, 'substantially a free gift to the city from Mr Donovan and hardly too much can be said in commendation of his public spirit'. The development was welcomed by all and the corporation contributed £750 towards the construction of the bridge.[232]

Thus it came about that at 11a.m. on Wednesday 7 March 1902, the new bridge, festooned with flags and bunting, was opened by His Royal Highness the Duke of Connaught, Commander of the Crown Forces in Ireland. Before his arrival, crowds gathered for the occasion. Students at the nearby university congregated on the bridge to the college grounds and amused themselves by whistling and singing songs, the most popular of which was 'The Boys of Wexford'. However, 'no marked demonstration of any kind occurred.'[233] The Duke of Connaught was accompanied by the 12th Lancers and members of the Royal Irish Constabulary on mounts and, upon his arrival, he was welcomed by the Earl of Bandon and the Lord Mayor, the Right Honourable Edward Fitzgerald, along with the developer, Thomas Donovan. 'The Duke's reception was of a respectful character and a slight cheer was raised as he alighted at the northern end of the bridge.'[234] As he made his way towards the bridge, a flag which was covering one of the inscription slabs was drawn by Mr W.H. Hill who then informed the Duke that the bridge was 'well and properly constructed'. The Duke was then invited to put into place the final stone of the bridge, following which he was presented with a silver trowel. On it was inscribed the message:

Presented to General His Royal Highness the Duke of Connaught,
Commanding the Forces in Ireland;
On the occasion of his opening Donovan's Bridge;
Built by Thomas Donovan, Cork,
May, 1902.

To cheers from the crowds, the Duke and the entire party of dignitaries crossed to the new buildings on the south side of the river, travelling via the new roads, one of which, Connaught Avenue, was named after the Royal Duke. The bridge itself is named to commemorate Thomas Donovan. A plaque on the wall says:

1902 Donovan's Bridge.
Presented to the citizens of Cork by Thomas Donovan, Fernhurst.
Right Honourable Edward Fitzgerald, Lord Mayor.
Augustine Roche Esq.High Sheriff.
W.H.Hill & Son Engineers.
Patrick Murray Builder.

The ceremony completed, the Duke left for Albert Quay and the terminus of the Cork Bandon and South Coast Railway from whence he departed for Bantry.[235]

Clontarf and Brian Boru Bridges

For many years there had been a belief among certain interested parties in Cork City and county that connecting the railways running south and west of Cork City from their termini on the city's south side, with the Great Southern and Western Railway at its terminus on Glanmire Road, would be of great benefit to the local economy. However, this was seen by others in the city, among them representatives of the trades and working classes, as detrimental to their interests as it could lead to the city being by-passed by rail traffic going directly to destinations outside of Cork.

Achievement of the project would necessitate bridges across both channels of the River Lee and it was deemed appropriate that plans for these bridges should be included for Parliamentary approval with other plans that would see a new bridge built across the River Suir at Waterford and the development of a rail link between Rosslare, Waterford and Cork. Following the enactment of a bill in 1898 to allow the development of railway services to Rosslare, in December 1899 a joint committee comprising members of the Cork Harbour Commissioners and Cork Corporation decided by ten votes to five that the city engineer and the Harbour Commissioner's engineer should inspect plans for proposed bridges across the two channels of the Lee, confer with the engineers of the Fishguard and Rosslare Railway Company and report to their respective boards.

However, when this decision was outlined to a full Harbour Commissioners meeting held on Wednesday 20 December 1899, some of the members stated that, 'bearing in mind that the erection of bridges across the Lee may have serious consequences for trade and other interests of the port and city, we see no reason to depart from the usual mode of procedure for having plans etc. laid before the Commissioners.'[236] Two points can be made here therefore: firstly, the fact that five members voted against

Brian Boru Bridge in open position in December 1911. (Courtesy Michael Lenihan.)

Brian Boru Bridge open for shipping.

the resolution at the sub-committee meeting substantiates that there was a body of opinion unhappy with the prospect of the bridges being thrown across the river channels and, furthermore, it can be said that these reservations pertained to certain trade or commercial interests.

The consequence of this situation was that little happened to advance the project for a number of years. Nevertheless, those in favour still actively pursued their ambitions, not least among them, interested parties in West Cork who stood to gain from a scheme that would enable the railways to have a direct through-connection from all parts of the country. In support of a bill before Parliament, in May 1905, a public meeting was arranged to be held in the city on behalf of the West Cork interests. This brought a response from the Cork United Trades Council which, while acknowledging the right of those from West Cork to fight their case, nevertheless stated unequivocally that:

> ... in the interests of the City of Cork and the working classes of the city it was necessary that the terminus of the Great Southern and Western Railway should remain in Cork and the bridging of the Lee, which would enable trains to travel to the western portion of the county, thereby depriving Cork of its many advantages, would be most injurious to the commercial and trading community of the city; and their council again endorsed the action of the corporation and the Harbour Board in opposition to the bill now before Parliament.[237]

Clontarf Bridge in the open position in December 1911. (Courtesy Michael Lenihan.)

Some days later a letter signed 'FO'C' was published in the local press suggesting that 'modern steam ferries' should be used to transfer goods between the different railways at a much lower cost than would be incurred through the bridge-building endeavours.

As well as the differences between these interested parties, disagreement also emerged between the representatives of the cities of Waterford and Cork over the distribution of monies for the proposed bridges in the two cities. In July, the Member of Parliament for Waterford, John Redmond, asked questions of the Chief Secretary regarding the proposals for a bridge across the River Suir and the status of monies which had been allocated for the Cork and Waterford Bridge projects. The reply stated that, 'representations had been made to me by various parties in favour of a Parliamentary grant in aid of railway connections in Cork as proposed by the Junction Railway Bill now in committee and of the bridge in Waterford'. He went on to say that the treasury had decided to:

> ... ask Parliament at a convenient opportunity to vote a sum of £93,000 in aid of the two proposals mentioned. It is intended to vote a sum of £60,000 as a contribution to the scheme for the junction at Cork ... the balance of the vote, £33,000, will be made available for the bridge at Waterford ...[238]

Clontarf Bridge no longer carrying trains or opening for shipping.

Two days later, the Cork United Trades Council unanimously passed a resolution protesting against the re-committal of the Cork Junction Railway Bill as, 'it would be detrimental to the interests of the commercial and industrial community of the city.'[239] The Trades Council was delighted that not everybody in Waterford was happy with the situation either.

Throughout the remainder of 1905, the corporation and the Harbour Commissioners were engaged in examining a variety of proposals pertaining to the planned crossings of the river channels. Then, in March 1906, a delegation from Cork met with the Secretary to the Treasury, Mr McKenna of Monmouthshire. Seeking release of the money for the bridges project, they outlined to the Secretary that this had first been made available by an agreement contained in the first schedule of the Fishguard and Rosslare Railways and Harbours Act of 1898, but withdrawn in 1901 and then restored by the treasury just a year earlier.[240] They reminded the Secretary of the arrangement whereby £60,000 was to be appropriated on behalf of the railway connection across the Lee and the remaining £33,000 to go to Waterford for a similar purpose. Mr McKenna, it was reported, said that as soon as suitable schemes were put forward, the money would be applied in that proportion.[241]

Two weeks later, however, having been visited by the members for Waterford, Mr McKenna said that while his earlier statement had been reported substantially correctly in the newspapers, no promises had been made by him. The bill for the works to be undertaken in Cork that would be paid for by the £60,000 was known as the Cork City Railways and Works Bill and this was now opposed by the Waterford members because of the money distribution. They argued that the whole of the money should be devoted to County Waterford purposes.[242] In a debate on 1 May 1906, on the second reading of the Cork City Railways and Works Bill, Mr J.J. O'Shee:

> ... strongly protested against any part of this money being used for constructing link lines or bridges across the Lee. The Cork City Railways Bill was really for the construction of a tramway. He had only risen to enter his protest against the £60,000 being allocated directly or indirectly to the construction of these works in the City of Cork, and to claim on behalf of the ratepayers of Waterford that when the allocation came to be considered the House would determine that as far as possible the money should be used in the way he had indicated.[243]

In reply, William O'Brien of Cork said that, 'by bridging the Lee, it offered the best prospect of bringing into the railway connection a vast district in the west which was now absolutely cut off from the rest of the country.'[244] As a result of this dispute, Messrs O'Brien and Roche of Cork blocked the bills for developments in Waterford while O'Shee of Waterford continued to block the Cork bills.[245]

On 10 May 1906, a Parliamentary enquiry opened into the merits of the schemes proposed for Cork, comprising chairman Mr Ashton, Mr Gay Baring, Mr Essex and

Irishman Mr Hugh Law. Three different Bill proposals were under consideration, eventually resulting in the Cork City Railways and Works Bill going before Parliament in May and June 1906. The Waterford members, however, continued to oppose the plans and when the Bill came to have its third reading in June, Mr O'Shee moved its rejection. The Financial Secretary Mr McKenna, however, reminded the Honourable Gentlemen, 'that this subject had been fully debated at various stages and that after every argument that could be used had been heard, the House was in favour of the Bill by 196 votes to 27 against. After such a crushing defeat he should have thought that the Honourable Member would not have the courage to discuss the case again'. After this the question was put again and passed.[246]

On 4 August 1906, the Cork City Railways and Works Act was given royal assent and provided for the building of railway sidings on the quays which the corporation had been developing since the middle of the previous century. In order that a link could be operated between the sidings and the national network at the Great Southern and Western terminus and the West Cork Railway terminus and Albert Quay, it would be necessary to consider shipping requirements. Cargo ships still operated right up as far as St Patrick's Bridge on the North Channel and beyond Parnell Bridge on the south. It was for this reason that new bridges on both channels would have to allow shipping through and there was much discussion as to the types of bridges that should be built.

Second in the series of four photographs taken by the GSWR in 1911.

Suggestions ranged from the centre-arch moving bridge like Parnell Bridge, to the old-fashioned portcullis. Eventually, however, the design decided upon was the Scherzer Rolling Lift Bascule Bridge, consisting of four sections resting on steel piles with concrete overlay. Concrete abutments, as in an ordinary bridge, were also part of the design. The term 'bascule' comes from the French word meaning seesaw or balance and this is precisely what the bascule rolling lift bridge did; pivoting on a roller, the bridge superstructure itself was the weight on one side with a counter-weight in the form of a box on the other. This specific bridge-type was the invention of William Scherzer of Chicago in the early 1890s, to take an elevated railroad over the Chicago river. The advantages of this type of bridge were threefold. Firstly the opening span was self-contained and for that reason would not take up any quayside space; secondly, the actual opening of the centre span was relatively fast, and finally the centre passageway for the shipping was free from piers.

In November 1909, the contract for the building of the bridges was awarded to the celebrated bridge builders William Arrol & Co., one of the largest bridge-building company in the world at the time. The company was founded by the Scottish-born civil engineer in the 1870s and among the more famous of his constructions were the Tay and Forth Bridges in Scotland as well as Tower Bridge in London. Supervising engineers for the Cork project were William Burnside on behalf of William Arrol & Co. and William Armstrong for the railways.

The construction and installation of the bridges went ahead with the sum of £15,000 being contributed towards the overall cost by the Cork Bandon and South Coast Railways. Brian Boru on the North Channel was to be 232ft long and Clontarf on the South Channel 197ft. Each would have an opening span of 62ft and roadways constructed of planks of hardwood laid diagonally.[247] Elements of the rolling lift bridge were supplied by the Cleveland Bridge and Engineering Company of Darlington in England.[248] The bridges, upon completion, would also allow vehicular traffic to cross the river and two new streets were to be built, Brian Boru at the North Channel and another adjacent to Dean Street on the south.

The bridges were constructed in an upright position; it was only upon completion that they were lowered into the horizontal. By April 1911, the skeletal lifting parts of both bridges were in place and the counterweights on the North Channel were installed. Preliminary work on the installation of the lifting gear was also underway. This included powerful electric motors that would lift each bridge against a wind pressure of twenty pounds per square foot in three and a half minutes. With no wind pressure that time would be reduced to two minutes.[249] Manually-operated emergency gear was also provided for each of the bridges.

Less than two months later, on 21 and 31 May respectively, the completed lifting sections of the North- and South-Channel bridges were lowered into place successfully for the first time.[250]

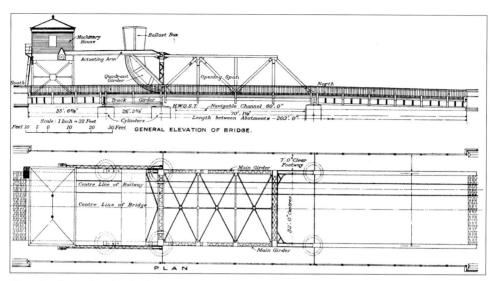

Plans for Clontarf Bridge for the purpose of repairs in the early 1930s. (Courtesy Michael Lenihan.)

In the lead up to the official opening of the bridges, during the month of December 1911, both were tested with fully laden goods trains. One test took place on Thursday 14 December with a final test being carried out on Friday 30. This final test was attended by Colonel Van Dunlop, chief engineer of the Board of Trade and on this occasion, two 50-ton locomotives, as well as a corporation steam-roller and a traction engine, were driven onto the bridge, while engineers observed the undersides of the structures for any possible deflections.[251] Satisfactory testing completed, the bridges were opened for business on 1 January 1912. The first goods train left from the Great Southern and Western terminus and traversed the bridges at 8.15a.m. Being the first day of the year, not many people witnessed the inaugural crossing, but as the day wore on and further scheduled trains crossed the bridges, crowds assembled in greater numbers.

Comment continued on the merits and demerits of the project by those who saw the bridges as detrimental to aspects of Cork trade and commerce – not least that a number of ferry-boat operators were now out of business – and those who saw the linking of the railways as beneficial to the city and county.[252] In the autumn of 1927, however, owing to losses incurred by the railway operations, plans were announced for the abandonment of the bridges by the railway companies, which were resisted by many in the civic authorities. The issue ran into the spring and early summer of 1928 when the railways relented on their plans and continued with their operations.[253]

Now the furthest downstream crossings on the river channels, their red colour was quite a sight for both traveller on land – whether by rail or car – and on water. Small timber engine-rooms were built on elevated platforms to control the operations of the bridges and the equipment installed was both electric and hand-operated.

The position of the bridge operator was an important one in the city, in that they always had to be on call and aware of the shipping movements in and out of the port. One such operator was Paddy Herbert, a native of Newcastlewest in County Limerick and a man who lived for many years in Carrigrohane. He also worked as a watchmaker in the city and would regularly have to leave his bench of tiny screws and mechanical parts and turn his attentions to the well-oiled mechanisms of the bridges.

Throughout the fifty-odd years that the lifting operations functioned, there was always great excitement when 'the bridges were up', particularly when each new generation of youngsters discovered it for themselves. Many people fondly recall journeying to the factories of Ford's or Dunlop's on bicycles and frantically trying to make it across the bridges before they were closed to traffic because of an approaching ship. The price of failure was perhaps being late for work.

By the early 1950s, both of these bridges had ceased to admit shipping and some forty-two years after they were first opened on 1 January 1912, in 1954 Brian Boru Bridge was revamped. Three years later, in 1957, Clontarf Bridge had to undergo the same treatment.[254] With the termination of the lifting operations, at least one group of the Cork public lost a valuable asset to their existence. The school-going population could no longer claim when late for school that 'the bridge was up sir'.

As well as being revamped, the bridges had lost their red colour, being painted a dull grey instead. On the afternoon of Wednesday 3 November 1965, that red colour could have indicated danger, for at 2.50p.m. a 1,200-ton-cargo ship, the *City of Cork*, moved up the South Channel towards her berth. Clontarf Bridge was not open; still the ship pressed on. People in cars on the bridge saw the ship and assumed that she must soon begin to slow; still she pressed on, over-running her berthing spot and heading towards the bridge. People abandoned their cars and dock workers looked on in amazement as the *City of Cork* ploughed into Clontarf Bridge to a distance of 8ft. Fortunately nobody was injured.

The ship was owned by Palgrave Murphy & Co. and operated on the Antwerp – Hamburg – Dublin – Cork run carrying general cargo. The captain was E.J. Gibbons and he had let out the anchor to slow his vessel – to no avail. The immediate effect of the accident was the forced introduction of a one-way traffic system in the city. Following an engineering investigation, it was found that one main supporting girder was fractured and needed replacing before the bridge could be reopened. Initially it was thought that at least a month would be required to effect repairs, but, within a week, a supporting girder, weighing 12 tons from a disused railway bridge in Desert, West Cork, was located and transported to the damaged bridge. This allowed the bridge to function until permanent repairs could be made and on 9 November the bridge was tested with the full weight of a steam train crossing it and was declared fully operational again.

In September 1976, CIE, which operated the railways systems throughout the country, decided to close the Cork Link rail service that had begun its cross-river operations in 1912. By the end of the month, the last trains traversed the bridges and over the following year, tracks on either side of the bridges were removed. In 1980, the lifting machinery for the Brian Boru Bridge was removed and the structure became a fixed one for pedestrian and road traffic. The following year, 1981, Clontarf Bridge underwent the same process.[255] This work was done by Bowen and Mulally, to designs by engineer Joe Shinkwin in collaboration with city engineer Sean McCarthy.

A strange incident occurred early in the 1980s connecting Brian Boru Bridge with that of St Patrick's Bridge, just upstream from it, in a way that perhaps was not the most desirable. Shortly after dark an emergency call was received at the Fire Station on Sullivan's Quay, to say that a trailer-load of hay had caught fire on St Patrick's Bridge. Under the command of Officer Pat Poland, a fire-crew headed immediately for the incident in their appliance. As they made their way along the South Mall and Grand Parade, however, they received a call from radio control asking them to divert to Brian Boru Bridge. Enterprising individuals on St Patrick's Bridge had managed to tip the burning hay-bales into the river, assuming they would be quenched in the waters below. However, the river, with a high tide, had just turned and the ebb flow now took the blazing hay-bales towards Brian Boru Bridge, beneath which they lodged and soon the timbers that formed the lowest element of the road surface were subjected to the full inferno rising from the waters.

The fire crew passed through St Patrick's Street and turned onto Merchant's Quay. There, the fire appliance over-heated and, with steam rising from the engine, Pat Poland ordered his men to take the hoses and make haste to Brian Boru Bridge to tackle the rising flames. As they ran towards the bridge, they met a bus inspector and others running the other way from the bus station located near the bridge. Asking them why they were running they enquired of the fire crew whether or not they were aware that beneath the bridge, in the path of the rising flames, ran the gas main to the south side of the city. The fire crews soon got their hoses turned on the hay, the flames were extinguished and life on Cork's bridges could return to normal.

War Damage: 1922

Throughout Ireland, the decade following 1912 was one of increasing tension and turmoil and Cork was no exception. The 1916 Rising saw the destruction of much of the centre of Dublin while 1920 was the year in which Cork became one of the most violent places in the country. At the end of that year, much of the city centre was destroyed on the night of 11 December in what became known as 'The Burning of Cork'. Over 240 yards of St Patrick's Street as well as the City Hall and the Carneige Library was burned by British Auxiliary forces.

Following on from independence in 1922, the pro- and anti-treaty forces engaged in a bitter civil war and so, in these difficult times, it was inevitable that further damage would be inflicted on the infrastructure of the country. Throughout large areas and in particular in Cork County, bridges were targeted by the Republican anti-treaty forces aiming to upset the communications and transport of the Free State troops. Headlines such as, 'Cutting Communications: Bridges Destroyed' and, 'More Destruction: Bridges Destroyed near Cork' were common in the newspapers.[256] One of the most serious of these was the destruction of the railway bridge at Mallow.

At the outset of the Civil War, the City of Cork was largely in the hands of the anti-treaty troops. When the Free-State authorities considered how they could take the city, they decided on an assault from sea, coming up Cork Harbour, landing in the vicinity of Passage and fighting their way up the estuary and into the city. This operation was undertaken in early August 1922 and, despite an attempt to block the river channel, within forty-eight hours of the operation getting under way the republicans were in retreat in all places. Inevitably, they sought to disrupt as they retreated and early targets included damaging the printing presses of the local *Cork Examiner* and *Cork Constitution* newspapers to prevent reporting of matters. As well as this, a number of attempts were made to inflict damage on city bridges to slow the Free State advance. Under a headline of 'Capture of the Rebel City', the *Freeman's Journal* reported that, 'the following bridges were destroyed by the irregulars – Brian Boru Bridge, Parnell Bridge and Parliament Bridge'.[257] A similar story appeared in *The Irish Times* declaring that 'Troops Enter Cork City' and that 'the following bridges were destroyed by the irregulars – Brian Boru Bridge, Parnell Bridge and Parliament Bridge'.[258] It is interesting to note that the wording of the bridge destruction reports is identical in both papers, obviously emanating from a single source. The next edition of the *Irish Times*, on Monday 14 August, gave greater detail regarding the effects on the bridges:

> Parnell Bridge, which spans the southern branch of the Lee ... on the direct route of the national troops into the city ... was the first of the bridges to engage the attention of the wreckers. After a terrible explosion a large hole was rent in the roadway while some of the iron-work was damaged.

The report went on to say that, 'Brian Boru Bridge, which spans the Lee a hundred yards eastwards of Parnell Bridge was next blown up'. A brief account of the history of the railway bridges followed and then the report continued 'Parliament Bridge, which was also blown up ...'

In the absence of local newspaper accounts emanating from Cork, the impression given by the Dublin reports was that the bridges were totally destroyed in the battle for the southern capital. There was indeed damage caused to the structures but it was not in any way as severe as the total destruction suggested in the reports. As soon as the *Cork Examiner* re-opened for business, on Wednesday 15 August, beneath a headline of 'Injury to Parnell Bridge' and a picture, the paper reported that an attempt was made to destroy the bridge with explosives, 'but it only twisted some of the stanchions and loosened a few boards on the foot-track'.[259] No mention was made on this occasion of damage to the other two bridges.

Three days later, the paper reported that an ambush took place in which, 'a bomb was thrown on Parliament Bridge at a lorry containing about twenty ex-servicemen recently joined the National army ...'[260] Clearly then, the bridge was operational at some level although on these, and indeed on a number of other occasions, damage was inflicted on the bridge during the fighting at this time. Repairs to Parliament Bridge would not be completed for a number of years as has already been seen.[261]

Daly's Bridge

Probably the most beautiful stretch of river within the confines of the city is that which flows between the gardens of the Sunday's Well Road houses and Fitzgerald's Park. There is always a great serenity and calmness about it and for many years during the nineteenth century and the early part of the twentieth, ferry boats crossed this part of the river. The operation was known as Dooley's Ferry, but despite the romantic nature of the boats plying back and forth on the picturesque waters, by the advent of the twentieth century it was believed that a bridge would serve the north-western quarter of the city better.

Early in 1910, a deputation consisting of Revd John Russell, C.C. Cathedral, Arch-Deacon Powell, Aldermen Dale and O'Connor and a number of others, among whom was businessman James Daly, waited on the Public Works Committee of the corporation and 'advanced the great advisability' of placing a footbridge at the site. They argued that an area of three and a half square miles of the south-western and north-western parts of the city would benefit. It was resolved that a sub-committee of councillors along with the city engineer, J.F. Delaney, would examine the proposal.[262] At noon on 25 April, Aldermen Mullaney and Stack as well as Councillors O'Keefe, Fitzgerald and Newman and accompanied by J.F. Delaney visited the site following which the engineer undertook to do a detailed report.[263]

This report came up for consideration at a meeting of the Public Works Committee on Wednesday 15 June 1910, but before any serious discussion could even begin, Alderman Stack proposed that the entire project be postponed for twelve months given the nature of the city's finances at the time, citing that the mortgage debt was now over £300,000. Furthermore, he argued, the ferry rights would have to be bought from Miss Dooley and the ground-rents for the access properties on either side of the river would need to be acquired from Lord Cork.

City Engineer J.F. Delaney then outlined his report, including reference to the need for stepped access from the main Sunday's Well Road. He also responded to questions regarding the impact any flooding may have on the project. The differing views held by those opposed to and those in favour of the bridge became heated. At one point Mr Delea suggested that the bridge was being built for, 'the swells of Sunday's Well and not for the working classes.'[264] Some consideration was also given to the cost of the project and a further submission was made in favour of the bridge by solicitor Mr Blake and M.J. Stapleton who said, in response to a report in the press that the bridge would cost £1,700, that they would not wish the corporation to go to that much expense but that they had ascertained that a structure similar in size and width to their requirements was to be put over the Derwent for £235 and they would be perfectly satisfied with such a structure. The mood of the meeting was not in favour of the project, however, and the final outcome was a postponement of the bridge plans for a period of six months.

It was not six months until the bridge project resurfaced, nor even six years, but the best part of sixteen years, during which time change on an almost unimaginable scale had occurred in Ireland and throughout Europe. The centres of both Dublin and Cork had been decimated through fire during the fight for independence and were now rising again from the ashes.[265] Twenty-six of Ireland's thirty-two counties now constituted the Free State of Ireland, governed by her own Dáil of the people's choosing. That government set about the business of nation-building, restoring law and order and demonstrating to the rest of the world that Ireland was a righteous and upstanding nation, equal to any in the civilised world. Law and order was paramount and perceived corruption not to be tolerated. Local government came under close scrutiny and in March 1924, following an inquiry into the affairs of the capital, Dublin Corporation was dissolved.[266]

Daly's Bridge more commonly known as the Shakey Bridge. (Courtesy Michael Lenihan.)

Postcard view of Donovan's Bridge with the entrance bridge to UCC also. (Courtesy Michael Lenihan.)

Prior to this, a body in Cork called the Cork Progressive Association (CPA) had been formed in 1923. Amongst its aims was municipal reform and having called on the Minister for Local Government to hold an inquiry into civic affairs in Cork, following the actions taken in Dublin, the CPA increased calls for the Cork situation to be examined. An inquiry was held in the autumn of 1924 which lasted for nine days, hearing submissions from members of the CPA as well as City Hall officials and Lord Mayor Sean French.[267] The outcome was the dissolution of Cork Corporation on 31 October and Philip Monahan was appointed Commissioner to manage all civic affairs by Ministerial order.[268] For almost all of the remainder of the decade, he took control of every aspect of civic administration in the city, including the work of the committees such as that of the Public Works.

On Thursday 30 April 1925, residents from the Sunday's Well area urged Commissioner Monahan to build the footbridge that they had sought many years earlier. Monahan's reply was that if the bridge was to cost but £300-£400 he might consider it but not if it ran to thousands. He added that the North Gate and Carroll's Quay Bridges were in a dangerous state and he couldn't take on another bridge project while this was the case. However, he agreed that he wouldn't make any final decision until after the city engineer had submitted a report to him.[269]

For those who desired to see the bridge built, however, help was at hand. On 28 July 1926, while transacting the business of the corporation standing committees,

Philip Monahan announced that four tenders for the erection of a suspension bridge at Ferry Walk had been received and that the lowest was from David Rowell and Co. from Westminster at £924. Without elaborating, he said that, subject to some other arrangements being finalised, he would accept this tender. S.W. Farrington, the city engineer, said that this price did not include the laying of foundations which would have to be borne by the rate-payers. The *Cork Examiner* speculated that 'a large portion of the erection expenses will be borne by a private Cork enterprise interested in the scheme'.[270] Two weeks later Philip Monahan publicly thanked, on behalf of the city, Mr James Daly of Dalymount, Sunday's Well for his generosity in presenting to the city the ferry rights at Sunday's Well which he had acquired and his even greater generosity in providing half of the cost of the erection of a suspension bridge over the Lee at this point.[271]

The first phase of the operation to construct the bridge was the building of the foundations and the access walkways to the crossing point. When these were completed, it was announced on 2 February 1927 by Farrington that the steel erectors would be coming on site that week to begin work on the structure itself.[272] On 22 February, the *Cork Examiner* reported that the main structure was completed; that the walkway would be finished that day; and all that remained to be done was the painting, which would be completed by the end of March.[273] The report again thanked James Daly and described how the bridge had been 'a hobby' of his for many years.

On Saturday 9 April 1927, 'a neat suspension bridge connecting the Sunday's Well side with the southern banks of the Lee at Ferry Walk was opened to the public' by Mr M. O'Driscoll.[274] Also present at the opening ceremony were members of the local Vincentian community from St Vincent's church, located not far from the bridge, and many city dignitaries. All who spoke paid tribute to James Daly, many saying that although originally and still a Ballyduff man, he had given with great generosity to Sunday's Well and indeed Cork. In turn, James Daly thanked all for their flattering comments and he particularly thanked City Engineer Farrington for the beautiful design plans that could be reasonable achieved.[275]

Thus, Daly's Bridge that gracefully spans the river today was built and opened. Suspended at points across the river from cables which are anchored to supports on either side, the bridge has a pedestrian walkway 4½ft wide made from timber planking. Daly's or the Shakey Bridge as it is known colloquially, has, perhaps more than any other bridge in the city, stood up to the pressures placed upon it by the populace. A short route to the playing fields of the Mardyke, Sunday after Sunday, hundreds of people would make their way across the bridge *en route* to soccer or Gaelic matches. As throw-in time neared, the shuffling of the masses would cause the bridge to shake, thus occasioning its nickname.

An unusual view of the Shakey Bridge. (Courtesy Michael Lenihan.)

The Latest North Gate Bridge

The next major bridge project undertaken was the reconstruction, yet again, of the North Gate Bridge. This was forced on the corporation when, in 1954, the bridge which Jerome J. Collins had built in the 1860s was condemned by city engineer S.W. Farrington who found fracturing of the corbels.[276] As a consequence, he declared the bridge inadequate for the traffic that it carried. Prolonged continued use of the bridge would only lead to increased danger and the corporation had no choice but to accept the recommendation that the bridge be replaced.

W.J.L. O'Connell was employed as the engineer on the project and it was on his advice that the present structure was built. Following surveys to locate rock for the sinking of new foundations, a new pre-stressed concrete structure was to be built, the first time that this material would be used in bridge construction in the city.[277]

The closure of the iron structure that had spanned the river at this point for all of living memory caused much sadness in the north-side communities. For these people, the nostalgic recollections of childhood trips to the city, of lingering kisses to a loved one, of everyday hustle and bustle on the bridge, could in no way compensate for the loss of such an aesthetically beautiful structure.

A leased Bailey Bridge was put into place just upstream from the construction site to cater for the needs of the population, a population that, having said goodbye to the old bridge, welcomed the new one on 6 November 1961. The Lord Mayor was Anthony Barry TD and he performed the opening ceremony, proceeding to the new bridge with the other councillors, all dressed in their official robes. Accompanying him also was the parish priest of St Peter and Paul's church, Very Revd Canon J. Fehily, who blessed the structure in Latin before the opening. (By the time another such blessing was needed, Vatican 2 had changed Church rituals and introduced the use of everyday spoken language.) The new Griffith's Bridge, as it was named at the ceremony in honour of Arthur Griffith, was 62ft wide, 22ft more than its predecessor had been. The pre-stressed concrete was in the form of a continuous 16ins deep slab, and the bridge had a fall of 2ft from north to south; its length was 108ft.

It was a great relief to the traders of the North Main Street when the bridge was finally opened, as with the closure of the 1860s structure, they had experienced a considerable falling off in business. Quite simply, many north side people didn't want to walk upriver to the Bailey Bridge and then down the other side before beginning their ascent of Shandon Street.

Griffith's Bridge will always be referred to as simply the North Gate Bridge. When opening it, Anthony Barry was opening at least the sixth bridge at the site – there may have been more but no records exist giving the exact number of bridges on this site during the Viking and early-Norman periods. The ribbon that he cut led into the past as well as the future.

The temporary Bailey's Bridge continued in operation for many years after 1961, having been bought by the corporation. In January 1991, however, it was dismantled and is now just another memory for those who lived in Cork in the 1960s, '70s and '80s.

Parnell Bridge Again

While no actual new construction occurred between 1926 and 1957, the corporation had been considering the state of the city's bridges. In 1946, Government approval had been sought for the replacement of Parnell Bridge; it had taken this long for officialdom to accept the cries of the population first heard within a week of the bridge's opening in 1882 that it was too narrow and generally inadequate. Despite the application, it was to be more than twenty years before anything definite would be done.

The question of the reliability of the seventy-two-year-old bridge was raised in 1954 when a public enquiry was held, eventually being adjourned until a report on current and future traffic requirements could be produced. On 8 June 1955, Cork TD's Pa McGrath and Jack Lynch asked the Minister for Local Government in the Dáil for an update on the situation. Replying on behalf of the minister, Minister for Health

T.F. O'Higgins said that the inquiry into the application of the Cork Corporation for a bridge order for construction of a new bridge to replace Parnell Bridge had opened on 11 May 1954 and that he expected to be in a position to give a decision on the matter in the near future.[278] A year later, Pa McGrath again asked in the Dáil whether the minister was now in a position to state the result of the inquiry? Minister O'Donnell replied that:

> ... this is a highly technical matter. The bridge is a swivel bridge. There is a right of navigation there at the moment and it is a highly technical question as to whether the new bridge should be a swivel bridge or a permanent structure. I may tell the Deputy that the matter has not yet come before me. It is being examined by my technicians. It will require very, very careful examination. But I can assure the Deputy that, when it does come before me, I shall give a decision very quickly on it.[279]

Nine years were to pass before yet another public enquiry was held in 1965, but this was adjourned pending the outcome of a traffic report which had been carried out over the previous number of years.[280] Then, on Friday 10 January 1968, the city was shocked when City Engineer Sean McCarthy, closed the bridge in order to investigate a crack that had appeared in the road surface at its southern end. The time was 4.55p.m.; the city's traffic was thrown into chaos. Upon investigation, corrosion of the web plate of a cross girder was discovered and, in order to effect repairs, the bridge

Second Parnell Bridge.

would have to remain closed for ten days. (The web plate is the vertical or middle section in an H-type girder.) Traffic diversions were immediately put into place and a map of the new routeings was published in the *Cork Examiner* on the following day.[281]

The bridge, however, would never reopen. On Friday morning, 19 January 1968, the headline in the *Cork Examiner* told the story, 'Cork Bridge to Close for Good'. The previous evening, a meeting of the corporation was convened at short notice. In the chair was Alderman Pearse Wyse TD, Lord Mayor. City Engineer Sean McCarthy reported to the members:

> There are approximately forty transverse beams connected at each end of the two lattice girders. Each connection is formed by four bolts acting as hangers ... that is to say the transverse beam is hung from the lattice girder by means of the four bolts at each end. One of these bolts has been removed for inspection. It revealed serious corrosion ...[282]

The members of the chamber could not believe what they were hearing. Mr McCarthy continued, 'The corrosion is so serious that if only a small portion of the three hundred bolts involved are similarly affected, the bridge could be considered unsound ...'

The engineer's recommendation was that the bridge be replaced with a fixed structure and the inquiry that had been adjourned nearly three years previously was reconvened under the government inspector who had been chairing it, a man with the somewhat appropriate name of Mr B.J. Lee.

Dr McCarthy (not Sean McCarthy the City Engineer) was a traffic consultant appointed by the corporation and he recommended the replacement of five bridges in all: Parnell, Clontarf, Parliament, the South Gate and Clarke's Bridges. The city manager, however, reported the City Engineer as being happy that apart from Parnell Bridge, the others could meet the requirements placed upon them.

On 1 February 1968, the matter was again the subject of questions in the Dáil, when Mr Barrett asked if the minister, 'is now prepared to make a bridge order in respect of Parnell Bridge, Cork, without waiting for the result of a traffic survey, in view of the urgent nature of the problem which has arisen following the closure of the bridge?' In reply, Minister Boland said that, 'a letter has issued to the corporation from my department indicating that I have given instructions for the preparation of an order authorising the construction of a bridge to replace the existing bridge; with the letter is enclosed a draft of the bridge order for examination by the corporation.'[283]

Construction Work Begins

By early 1969, a temporary bailey bridge was placed across the river at Morrison's Island, just upstream of the construction site. This was to remain in place until the project was completed, an operation that took fifteen months in all, with the old bridge being completely demolished by November 1969. Two features of the old bridge were retained in the new structure, however: the abutments and two lamp standards.

On 24 May 1971, the new bridge was opened to the public by Lord Mayor Alderman Peter Barry TD. He was following in the footsteps of his father who had performed the previous opening ceremony of a bridge in the city, Griffith's Bridge at the North Gate ten years earlier. This was a unique situation in the history of the city as it was the only time that a father and son had each opened new bridges while holding the office of Lord Mayor. Furthermore, bridge engineer Barry O'Connell of O'Connell Harley Consulting Engineers, was the son of W.J.L. O'Connell who had designed Griffith's Bridge at the North Gate.[284] In the course of his speech, Lord Mayor Barry said that the new bridge was far superior to the one it was replacing, not least because of the fine views to be taken of City Hall on the one side and the Savings and Provincial Banks on the other. Having cut the tape and declared the new bridge open, the Lord Mayor and the Councillors walked across the bridge and then made their way to the Imperial Hotel for a celebration dinner.

The new Parnell Bridge is a very wide one, accommodating four lanes of traffic with an island in the centre and also having wide footpaths. A plaque on the South Mall side states:

Parnell Bridge. This bridge was opened on the 24th of May 1971 by Alderman Peter Barry TD, Lord Mayor of Cork. Finbarr O'Connell, Engineer. Eamonn Byrne, City Architect and Planning Officer, Cork Corporation.

Following this on the plaque is the city's coat of arms. On the Union Quay side of the bridge, the same appears on a plaque in Irish.

Thus ends a story that began in the late 1820s and undoubtedly it will be many more years before another bridge is needed at the site.

Trinity Bridge

The last bridge to be constructed before the 1980s was Trinity Bridge, a footbridge that spans the river between Morrison's Island and Union Quay and carrying the same name as the nearby Holy Trinity church, whose spire reaches skywards and towers above the bridge. Built in 1977, the bridge was opened by Lord Mayor Gerald Goldberg on 14 October in that year. With a 100ft span and a 12ft walkway, the bridge cost £42,000 to construct; but it wasn't just any £42,000. The money came completely from parking charges in the city, so that in the truest sense, the people paid for their own bridge. It was designed by Cyril Roche of O'Connell Harley on the South Mall and the construction was carried out by a company called Public Works Ltd.

Cyril Roche did not live to see the bridge opened but his widow Bridie attended the ceremony. Under the supervision of City Engineer Sean McCarthy, the bridge took twenty-four weeks to complete. At a reception following the opening ceremony, held in the nearby Imperial Hotel, Lord Mayor Goldberg said that the bridge was a symbol, not only because it bridges the river at a most important point, but it also brought to fruition one of the long-term plans of Cork Corporation in relation to the traffic situation in the city.[285]

Since its opening, it has served the people of the south side well; without it they would have to cross the river via Parnell or Parliament bridges. It is probably the only bridge in the city that in recent years has on occasions been virtually submerged by the river during severe flooding (the walkway, not the railings), and city Gardaí have had to carry people from the bridge as the floodwaters lapped their way over it. But this is rare and the convenience afforded by this bridge would have been much appreciated by those of generations past.

As the twentieth century progressed the nature and symbolism of the rituals associated with the opening of bridges in the city changed from that of previous centuries. No longer did Masonic Lodges process in full regalia through the city streets. Neither were wines poured on the key-stones nor elaborate scrolls and inscribed memorabilia placed inside the structures as testimony to those involved in the projects. Nevertheless, powerful businessmen and commercial interests combined with the civic authorities to achieve their ambitions and bridges such as Daly's were

Trinity Bridge with Trinity church spire in the background.

named in honour of benefactors to the city; religious leaders now intoned their blessings on new structures. Records of those involved were displayed more overtly in the form of inscribed wall-plaques.

Twentieth-century bridges differed aesthetically from earlier structures. Some reflected new developments in bridge design elsewhere in the world; others displayed the simple functionality required in a modern city.

The End of the Twentieth Century

From St Finbarr, the founder and patron of the City of Cork early in the seventh century, to the present day, Cork has grown and expanded. It began as a small monastic settlement on a hill just south of the river estuary. Now it spreads in all directions, filling the entire valley. The city has never really been in decline and has survived all that man and the elements have thrown at it. Nature's best efforts at subduing it was in 1622, when a lightning storm caused the entire city centre to catch fire. Fifteen hundred houses were destroyed and many hundreds of people lost their lives; the city, however, survived.[286] In 1690, the siege of the city was to cause great damage, but rather than bring Cork down, the effects of the siege served only to encourage further expansion beyond the old medieval walls onto the surrounding marshy islands. This, in turn, led to increased bridge-building projects as the islands were joined together to form the modern city.

De Valera Bridge from on high.

De Valera Bridge.

Throughout the nineteenth century, the nature and type of bridges that were built across the twin channels of the Lee that encompass the centre of Cork encouraged social and economic growth and the political masters of the day were centre-stage when ceremonial openings of the bridges occurred. The identity of the bridges that spanned the river to the Corn Market, as expressed through the naming conventions employed, reflected disputed political orthodoxies and a changing mood in Cork and indeed national politics. During the twentieth century the rituals and identities of a new state and its power-brokers were those incorporated in the bridge projects up to the 1970s.

Éamon de Valera and Michael Collins Bridges

Such was the rate of expansion in the Cork of the 1970s that the corporation was forced to make appropriate plans that would also cater for any future growth. In 1968, when Parnell Bridge was about to be replaced, the traffic expert Dr McCarthy predicted that such future plans would be necessary. In September 1978, a new report, the Land Utilisation and Transportation System (LUTS for short), was issued by the planning departments of both the city and the county. Consultant reports, drawn up as part of the plan, identified river crossings as being highly significant for future traffic requirements throughout the area and said that a major site for, 'a potentially significant crossing downstream of the city centre is Dunkettle to Mahon' and that 'a high bridge or a tunnel would be required here'.[287]

In order that the city traffic would flow more freely, the ring-road concept was decided upon as part of the plans. But this ring road would have to cross the two channels of the Lee. So plans were formulated for two bridges to be built downstream from the Clontarf and Brian Boru Bridges. The LUTS stated that these bridges would be built in phase two of the plan and it, 'proposed that the Custom House Bridge South should be constructed at the beginning of this phase for continuity of design and construction with the North Bridge'. Furthermore, it also said that the Opera House Bridge should be 'accorded a similar high priority'.[288]

The result was the opening, on 19 November 1984, of Éamon de Valera and Michael Collins Bridges, costing a total of £3 million and built by Ascon Ltd, a company whose slogan proclaimed 'Bridge Builders for the Nation since 1956'. The consulting engineer for the project was J.D. Shinkwin and the City Engineer was W.A. Fitzgerald.

Lord Mayor Liam Burke TD performed the opening ceremony and told the assembled crowds that the purpose of the bridges and other developments was, 'to cater for the industrial pragmatism of the future'. He highlighted the commercial losses and the gains that the Cork of recent years had had to endure, but said that, 'the horizon now brightened ...' Listening were two freemen of Cork, former Taoiseach Jack Lynch and music professor Aloys Fleischman. Also present at the opening ceremony were relatives of the two statesmen after whom the bridges were named: Prof. Éamon de Valera, son of the former president, with his wife Sally and Michael Collins from Waterford, nephew of the west Cork General.

As always, when the time came for names to be given to new developments in the city, there were those who objected to the final choices. The majority of the citizens, however, approved of naming the bridges after Éamon de Valera and Michael Collins. Councillor Jim Corr, chairman of the Roads Committee of the corporation, said at the opening ceremony that regardless of political affiliations, the corporation wished to acknowledge and perpetuate the memory of the part played by Michael Collins and Éamon de Valera in the shaping of modern Ireland. It had been seventy-two years since the city had received two new bridges that served both traffic and pedestrians and those two bridges had been named after the great Brian Boru and the place of his great victory over the Danes, Clontarf. 'Now, after long and careful consideration, Cork Corporation decided it would be most appropriate to dedicate the new bridges to the memories of perhaps the two most out-standing political figures of the twentieth century, Michael Collins and Éamon de Valera.'[289]

So it was that by symbolically bringing together leaders of the two factions of the Civil-War divide that had cast a shadow over Irish politics and society for over half a century, Cork took its first steps on the road to a better future. As mentioned in the earliest pages of this book, these were the final two bridges to be named with political affiliations, although a further three bridges would be built across the Lee before the century ended.

Nano Nagle Bridge

About six weeks after the opening of the Éamon de Valera and Michael Collins Bridges, a great year in the history of Cork began. The anniversary of the granting of a charter to the city by John, Lord of Ireland and subsequent King of England, was marked by the Cork 800. The accuracy of the date date of this celebration was challenged by some historians, which was met with apathy by some sections of the Cork public. Many celebrations, were held throughout the city during the course of the year and on 14 June a number of new developments

Nano Nagle Bridge, named after the foundress of the Presentation Order of Nuns.

were opened by Lord Mayor Liam Burke TD. These included the pedestrian walkway at Hibernian Way, crossing the South Link Road; a new amenities park at the foot of Shandon Street; and a new footbridge, Nano Nagle Bridge, crossing the South Channel of the Lee from the end of the Grand Parade to Sullivan's Quay.

This bridge is a single-arch structure built by a company called Site Services Ltd, with John O'Donovan as the consulting engineer. The cost of the project was £160,000 and it was fitting that it should be named to commemorate Nano Nagle, one of Cork's most famous daughters. Born during the penal times and despite the repression of those years, she founded the Presentation Order of nuns and set up the South Presentation Convent at 50 Cove Street in 1777. She died in 1784.

Nano Nagle Bridge is uniquely located in the city. To the west are the South Gate Bridge, St Finbarr's Cathedral, Proby's Bridge and many other monuments from Cork's past. To the east lies the thriving commerce of the South Mall, Parliament Bridge, the spire of Holy Trinity church and yet more of the beautiful City of Cork, old and new. In his address to the crowds at the opening ceremony of the first amenity (the Hibernian Road crossing), Lord Mayor Burke said that,'this year we have witnessed the re-birth of the city and re-discovered the pride in being Cork.'[290]

Christy Ring Bridge

As the Cork 800 year of 1985 closed, the corporation was working on plans for yet another new bridge on the North Channel of the Lee, plans that upon completion would leave the city's compliment of river bridges increased by four since the beginning of the decade.

For well over a century, the suggestion that a bridge be located at the end of Mulgrave Road and crossing over to the Emmet Place had been heard in the city. Once it had been the north side merchants arguing for it – now it was the driver who sat frustrated in his car with traffic backed up over St Patrick's Bridge and down the length of St Patrick's Street. By 1987 this bridge would finally be built.

A Century Earlier
At a meeting of the Improvement Department of the corporation on Friday, 10 July 1863, a deputation waited on the members, their objective being to seek the aid of the corporation in building a footbridge across the river at the end of Mulgrave Road. It would, the delegation argued, be a very great boon to, 'a numerous and influential body of rate-payers, the butter buyers and merchants of Cork.'[291] Then, at the corporation meeting of Monday 30 November, the Mayor asked for a report on the possibility of relocating the 1858 wooden bridge that lay to the west of the North Gate Bridge at the Mulgrave Road site.[292] This action was recommended by clerk of works at the North Gate, Jerome Collins, in a letter to the Improvement Department on 4 March 1864.[293]

During a discussion on the matter, Councillor Hegarty thought that the bridge should be placed halfway between the North Gate Bridge and St Patrick's Bridge, just above St Mary's church. This suggestion prompted a letter to the *Cork Examiner* a week later from Mr D.F. Leahy of Chapel Street saying that if such a project went ahead, it would cost the corporation very little because, among other things, 'the several brewers, distillers, farmers etc. having concerns in Blackpool, have already consented to subscribe.'[294]

On the same day, another delegation attended the Improvement Department meeting seeking a bridge at Mulgrave Road.[295] Six months later, a further delegation attended the corporation meeting of Monday 29 August, again seeking the Mulgrave Road Bridge. Councillor Mullane said, however, that this matter had been before the members on two previous occasions and had, on each of these, been referred to committees where it had been rejected.[296] The matter then went into abeyance, but only for a short number of years.

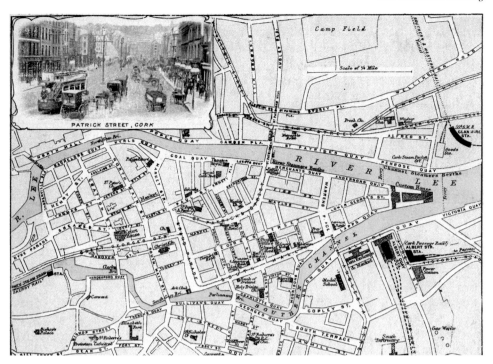

Postcard of tram routes including bridges. (Courtesy Michael Lenihan.)

At a meeting in July 1870, 'Mr Sheehan, in pursuance of notice, moved that in the opinion of this council it is desirable and expedient that a footbridge should be erected across the North Channel connecting Half Moon Street with Mulgrave Road and that Sir John Benson be requested to design plans for it and prepare an estimate of cost'. The motion was opposed by Mr Hegarty who said that the issue had been dealt with before and there was no reason to revisit it. Following a lengthy discussion, the motion was defeated by seventeen votes to eleven with two councillors not voting.[297] Thus it was that in the 1980s, when plans were agreed for a bridge to be built at the Opera House location, the aspirations that some councillors and lobby groups had had over a century earlier were being fulfilled.

Project Underway
The project got underway with the placement across the North Channel to help alleviate the traffic situation, of the Bailey Bridge that had once stood at Morrison's Island while the Parnell Bridge was being built. Following this, the construction of the permanent bridge got underway. Ascon Ltd was again the contractor and having completed the job in eighteen months, the bridge was officially opened by Lord Mayor Jerry O'Sullivan on Friday, 13 February 1987. It was named Christy Ring Bridge and Lord Mayor O'Sullivan told the dignitaries and people of Cork that,

'Christy Ring was the classic example of fighting back to win', a reference to the difficult economic situation prevailing at the time. 'Let us as Cork men and women unite as a team and try to follow his example,' he said. Christy, of course, was one of the most famous exponents of Ireland's national sport ever to grace a hurling pitch. One of those in attendance to hear his praises sung was his widow Rita.

Referring to the bridge that he now declared open, Lord Mayor Jerry O'Sullivan said that, 'it was in sharp contrast to the classic outlines of St Patrick's Bridge. It effectively symbolised the modern City of Cork with its bustling traffic lanes as distinct from the more leisurely era of bygone days when horse carriages and trams were the means of transport.'[298] The bridge cost £7 million and City Manager Jack Higgins as well as City Engineer Liam Fitzgerald, were both fulsome in their praise of everyone involved in the project.[299] To coincide with the opening of the bridge, a newly installed fountain was switched on in Emmet Place, just near the entrance to the Opera House. To mark the Year of the Environment, the fountain had the distance from Cork to all the capitals of the European Economic Union states inscribed on its base.

The exact location of this bridge is significant in that, as the newest of Cork's river crossings, it spanned the Lee at a point where, in olden times, travellers crossed *en route* to the walled city from the north-eastern reaches of the county and beyond. It was at this spot on the river that Munchgaar and the troops of the Duke of Wirtenberg had crossed and from whence they led the assault on the besieged city's eastern wall in the Williamite campaign of 1690.[300]

It is significant to note that on the occasion of the opening of the Christy Ring Bridge, Lord Mayor O'Sullivan referred to both the symbolism and the aesthetics of the new bridge as representative of Cork at the latter end of the twentieth century. Ironically the bridge site itself was the point at which those who inflicted the damage on the old city walls that led to subsequent bridge-building endeavours had crossed *en route* to their victory at Cork and also the place about which much nineteenth-century debate had occurred as local interested parties sought to have a river-crossing built there. The 1987 structure therefore, stands not just as a symbol of twentieth-century Cork but also as a tribute to events and people that helped to shape Cork in previous centuries.

St Finbarr's Bridge

Following the economic recession of the 1980s and the early 1990s, with the economic up-turn of the final years of the century, the increasing wealth of the people was manifest in, among other things, the numbers of private cars and other vehicles. Consequently, the city's traffic was a matter under consideration by the authorities. One of the issues uppermost in the minds of the commuting public was the lack of parking facilities in the city centre. As the century came to a close, in conjunction with Dublin development firm Howard Holdings, the corporation oversaw the building of a new bridge, as well as two high-rise car parks to help in this situation. Costing £20 million in all, this project included the car parks which were built at Sharman Crawford Street and that to the rear of City Hall, providing an extra 730 car spaces in the city, as well as the bridge, which was named after the patron saint of the city, St Finbarr.

The new river crossing cost £1 million and connected Sharman Crawford Street with the Western Road, in the shadow of St Finbarr's Cathedral; it also provided access to one of the new car parks. The contractor was P.J. Hegarty & Sons, overseen by City Engineer Kevin Terry and the bridge was opened by Lord Mayor Damian Wallace on Monday 6 December 1999. In his address, after declaring the bridge open to the public, the Lord Mayor said that the new river crossing re-opened access to one of Cork's most ancient areas.[301]

The name of the bridge, St Finbarr's, was appropriate as it was in this area that the earliest settlers, followers of St Finbarr himself, developed Cork's first living area. One of those present at the opening was Howard Holdings director and former president of the Cork Chamber of Commerce, Frank Boland. In his speech, he said that his company had a significant interest in Cork and that Cork was a major growth area.[302]

More than eleven bridges were newly constructed or replaced across the twin channels of the Lee over the course of the twentieth century. Each was a testimony of the events and contexts that led to its developments. Access to new suburban housing; the joining of diverse urban and rural districts through the

St Finbarr's Bridge on the South Channel.

interconnection of railway systems; the alleviation of increasing traffic congestion; all were facilitated through the building of bridges. From single-arch limestone to Scherzer Rolling Lift Bascule to modern pre-stressed concrete structures, the variety and types of bridges that were built over the course of the twentieth century added to the shape and look of the city as it grew outwards from the island centre. The ceremonies associated with the opening of the bridges perhaps lacked the pomp and ritual of former times. Nevertheless, consideration reveals that society's power-brokers used the occasions to advance their beliefs and agendas.

Thus ended the twentieth century. In particular, the 1980s was a decade to rival those of the 1850s and 1860s in the annals of river bridge-building in Cork City. As the twenty-first century dawned, plans were already under way for yet more bridges across the twin channels of the Lee that create the island City of Cork.

6

THE TWENTY-FIRST CENTURY

... seemingly in the blink of an eye old buildings are disappearing and a new city is rising up and expanding all around us.[303]

When historians of the future reflect on the first decade of the twenty-first century, they will see that the bridges constructed in that decade were all pedestrian, designed with community and welfare initiatives in mind. Another manifestation of this vision for the future of the city was the redevelopment of St Patrick's Street, opened in 2004 and which was designed by Catalan architect Beth Gali. The initiative was intended to create a pedestrian-friendly space in the centre of the city and subsequent redevelopments of other city-centre thoroughfares, such as the Grand Parade, continued this initiative. The work on St Patrick's Street was completed just before the city assumed the mantle of European Capital of Culture, in 2005. A number of other projects were also developed with this prestigious occasion in mind. These included the building of two pedestrian bridges across the North Channel of the river, Shandon Bridge, which was also intended to mark the millennium and Mardyke Bridge.

Shandon Bridge

On more than one occasion during the 1850s and the 1860s, when a number of footbridges were in use while the St Patrick's and North Gate Bridges were being rebuilt and various delegations were suggesting that a bridge should be built between Mulgrave Road and Half Moon Street, there had also been suggestions that a bridge be built across the North Channel of the Lee between Pope's Quay and Corn Market Street.[304] Just like the Mulgrave Road suggestion, however, the proposed Pope's Quay bridge was not built at that time.

As the twenty-first century approached however, Cork Corporation held a competition, in conjunction with the Institute of Engineers in Ireland, for the design of a new bridge to be placed at Cornmarket Street in celebration of the millennium. In all thirty-seven entries were received from Ireland, the United Kingdom and Europe and the winning entry was that from McGarry Ní Éanaigh Architects in association with Muir Associates Consulting Engineers.[305]

The winning design featured two symmetrical arches as the dominant visual feature, in a light steel structure. Forty metres long by three wide, the bridge provided a direct and easy connection between the north and south quays of the river's North Channel. Because the bridge slopes evenly from north to south, with a gradient of one in fifty, it is, to all intents and purposes, flat and this is important in terms of disabled access. Two separate lighting features were incorporated as part of the bridge. The first was an ambient light, illuminating the pedestrian walkway and the second was to light up the arches. The anti-slip walkway was made from sustainable tropical hardwood. Construction of the bridge was by Ascon Ltd and the steel contractor was S.H. Structures Ltd of Sherburn in Elmet, near Leeds in West Yorkshire.

Although the bridge may have been intended as a millennium project, the opening did not occur until Thursday, 9 December 2004, just in time for the Capital of Culture year and was performed by Lord Mayor Sean Martin. Costing almost €2½ million,[306] a number of different names were suggested for the bridge, such as Millennium and Cornmarket, but the name Shandon Bridge was chosen. This was appropriate because the bridge connects the old area of Cornmarket Street with the Shandon area, via one of the many lanes that run from the north side of the city to the quaysides below.

Shandon Bridge with Shandon Steeple behind.

Mardyke Bridge

On 8 December 2003, the corporation adopted the *Cork City Development Plan, 2004,* which came into effect on 5 January 2004. Policy Chapter 10, section 125 of the plan outlined details of an initiative known as the Banks of the Lee Project which was undertaken by the corporation as part of the millennium initiatives for the city and also as part of planned developments to coincide with the city's reign as European Capital of Culture in 2005. This plan envisaged an amenity walkway from the North Mall immediately to the west of the North Gate Bridge on the northern bank of the North Channel - across Wyse's Bridge and through Distillery Island, then across a new bridge spanning the North Channel proper – westwards along the Mardyke – across the Western Road at Thomas Davis Bridge and on through the lands of the Sacred Heart church – across yet another new bridge spanning the South Channel of the river and finally on to the Lee Fields.

Policy S28, of the 2004 plan, outlined in more detail the phases of the development and proposed to regenerate, '… the Lee Fields as a regional and city park and one of the city's major recreational and amenity open spaces.' As far as bridges were concerned, the Mardyke was to be linked to the Lee Fields through the construction of a pedestrian bridge over the South Channel of the Lee while the Mardyke Walk and the Distillery Island would be joined through the construction of a pedestrian/cycle bridge across the North Channel of the river.[307]

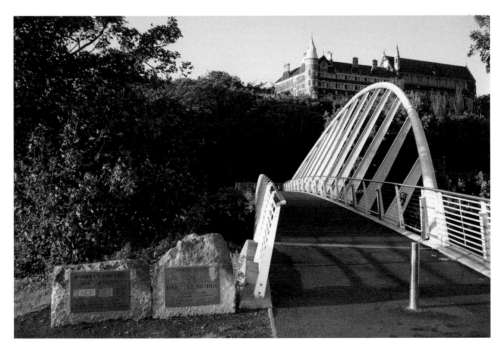

Mardyke Bridge on the Banks of the Lee Walk with St Vincent's church on the hills above.

On Monday, 28 February 2005, this bridge across the Lee's North Channel to connect the western end of the Distillery grounds with the Mardyke on the other side of the river was lifted into position. It was constructed by Harland & Wolff of Belfast, the famous ship-building company that built the ill-fated *Titanic* and the project took a total of thirty-four weeks to complete. The bridge was designed by Fehilly Timoney Gifford, consulting engineers and the project was undertaken with the active partnership of University College Cork who gained a cross-river access from the main campus to their department at Sunday's Well.[308] The contractor was John Flemming Construction and on Friday 17 June 2005 the Harland & Wolff constructed bridge across the North Channel of the Lee was opened by Lord Mayor Sean Martin.

The bridge is made of steel, has a span of 57m and its dominant feature is a single arch inclined at twenty-six degrees to the vertical. The walkway is 3m wide and the bridge is supported by abutments laid on precast concrete piles driven 10 to 15m below ground level. Before it was lifted into place in February 2005, a full trial erection was carried out in Belfast to ensure that all was in order. Following this, the bridge was transported to Cork by road. A 1,000-ton crane was required to lift the bridge into position and this was only done after some ground improvement works were carried out in order for the lifting operation to be completed. In all, 62 tons of steel were used in the bridge, which cost just less than €1 million to build.[309]

At a meeting of the corporation held on 11 April 2005, Councillor Dave McCarthy proposed that the new bridge on the Mardyke Walkway be named the Roy Keane Bridge. This suggestion, however, was not adopted and the bridge was officially named the Mardyke Bridge. On 1 February 2006, Lord Mayor Deirdre Clune, daughter of Peter Barry and grand-daughter of Anthony Barry, officially opened the North Mall to Mardyke or Slí Cumann na mBan section of the Banks of the Lee Walkway.

Cork's Newest River Bridges

Two further bridges have been built since the completion of the Shandon and Mardyke Bridges. While both of those cross the Lee's North Channel, the newest two bridges cross the river's South Channel. The first is known as the Western Link Bridge and is located to the rear of the University's Computer Science Complex in the Western Gate Building, on the site of the former city dog track and joining it to the Health Services Compex in the grounds of the old Brookfield House. Built by BAM construction, whot also built the new Cork Airport Terminal and carried out the refurbishment of the County Hall, it is a short pedestrian bridge, just downstream of the O'Neill Crowley Bridge and it is used mainly by students and university staff. The second, and at the time of writing the newest bridge across the Lee in the City of Cork, is that referred to in the *Cork City Development Plan 2004*, Chapter 10, section 125, which proposed the

Effigy on Lancaster Quay Bridge.

extension of the Banks of the Lee walkway, 'across a new pedestrian bridge linking to the walkway at the rear of the Kingsley hotel and onwards to the "Lee Fields"'.[310]

It will fall to future historians to evaluate the contexts and events that surrounded the building of bridges across the Lee during the first decade of the twenty-first century. A decade of contrasting economic fortunes, the planning and construction of the bridges took place during affluent times. That it was only pedestrian bridges that were constructed, coupled with the pedestrian-friendly designs of the reconstituted city streets, is suggestive of desire to achieve a more people-friendly society. The naming conventions employed, being apolitical and location oriented, suggests an enhanced pride of place. The incorporation of the newest bridges into an amenity walk, closely associated with the river in a broad sense emphasises a desire to refocus attention on the river in a number of ways, among them recreation and tourism. Finally, the simple elegance of design features such as arches and lighting can be compared with new bridges elsewhere such as the James Joyce and Sean O'Casey Bridges in Dublin, making Cork just one part of a wider movement for the betterment of infrastructural aesthetics.

It is worth noting that these elements are reflective of bridge developments much further afield than Dublin in this period. The tied-arch Zubizuri Bridge, also known as the Campo Volantin Bridge, spans the Nervion river in Bilbao in Spain. It was built in 1997 to the design of architect Santiago Calatrava. Its arched superstructure is

Lancaster Quay Bridge leading to twenty-first century developments.

Sun behind Lancaster Quay Bridge.

The new Cork and the old Lancaster Quay Bridge and apartments with St Finbarr's Cathedral behind.

clearly reminiscent of that on the Mardyke Bridge, while interestingly the see-through walkway is replicated in the James Joyce Bridge in Dublin.

Opened to the public in September 2001, the 126m Gateshead Millennium Bridge spanning the River Tyne in Newcastle in the north of England was the world's first tilting bridge. Again it has a distinctive arched superstructure and was constructed as part of a major pedestrian/cycle amenity in that city.

Finally, the Aagade Bridge in Copenhagen, built at a cost of €2 million between 2005 and 2006, was also part of a pedestrian/cycle route for that city. Reminiscent in appearance of the Shandon Bridge, amongst its noteworthy features are the elegantly lit handrails.

These are but three examples of bridge developments in other parts of Europe in the final years of the twentieth century and the first decade of the twenty-first. The points of similarity with the Irish bridges of the same period suggests that some form of international movement, both in terms of design and amenity provision, informed the context within which the Irish and indeed Cork developments occurred.

But was it not always so? The earliest bridges spanning the Lee at the North and South Gates of the walled city were drawbridges, largely for defensive purposes, which are typical of a walled city or town.

CONCLUSION

A Journey through time on the Bridges of Cork:
Twelve hundred years of spanning time …
Of Timber, Iron and Stone

This journey through the history of Cork's River Lee bridges is at an end. It is a journey that spans more than a thousand years and one that has touched on nearly all aspects of the city's development. Yet before we leave the pillars and arches of these bridges, it is worth taking a final look at them as they stand in majesty above the waters of the Lee and to see them in their geographical location in the city.

Banks of the Lee Walk footbridge joining the grounds of the Sacred Heart church with the Lee fields.

Arch reflection at entrance to UCC.

From the west the river comes, flowing gently beside the Carrigrohane Straight Road until it arrives beneath the towering edifice of the County Hall. There, after tumbling over a weir, it divides into two channels, one flowing to the north and one to the south of the city. Let us follow the South Channel, beneath the footbridge that brings the Slí na Laoi amenity walkway to the Lee Fields, through the grounds of the Sacred Heart Missionary Fathers, their house and church to the left, and on to the next bridge on the channel, the O'Neill Crowley Bridge of 1820. Flowing under this bridge, the river passes beneath large volumes of traffic travelling this western route from the city and enters into a much quieter part of its course. Just before a gentle bend to the east, this main South Channel is joined by the Curragheen River (actually a combination of the Curragheen and Glasheen Rivers), and flows on between the residential apartment and housing complex of Brookfield to the south and the new university computer complex to the north. Here the channel flows beneath the next bridge, a new pedestrian crossing joining the computer and medical faculties of the university and another that is reminiscent in appearance to the Shandon Bridge. Just downstream from here the river flows beneath two more pedestrian bridges joining the main Western Road with the Bon Secours Hospital and the University's Statistics and Mathematics department. These are nice bridges on which to pause perhaps for a moment of quiet reflection, the gently rippling

Entrance Bridge to UCC prior to its destruction in 1916. (Courtesy Michael Lenihan.)

waters beneath and the sound of birdsong in the nearby trees, all soothing to the ear and pleasing to behold.

The river travels onwards and next flows beneath Marc Isambard Brunel's Gaol Bridge of 1835. On the southern side of the bridge stands the Portico of the Old Gaol and just a little downstream, the original main entrance to the university campus from which, all the way to the college's modern entrance further downstream, is a pathway on which you can stroll in peace enjoying the atmosphere that only a river-walk can bring.

For thirty years, between the opening of the college in 1849 and 1879, the authorities had sought an entrance to the campus that was removed from the Gaol and nearer to the city. This was achieved in 1879 when, on the occasion of a visit to the college by the British Medical Association, the new entrance was opened to the public. At this main university gate stands Cork's Bridge of Academia and one we haven't yet met on our journey. The current bridge, across which thousands of students travel *en route* to lectures in the college each year, dates to the late 1920s. It is the third bridge at the site; the original 1879 bridge was made from Oregon pine but this was replaced in 1910 by one of ferro-concrete designed by James Walker with the assistance of Professor Alexander of the college's engineering department. It lasted only but six years, being destroyed by a flood in 1916. The replacement took another twelve years, not least because of wartime shortages. Built a short distance downstream from the site of the previous two bridges, a shop and houses were demolished to facilitate reconstrustion.

Entrance Bridge to UCC. (Courtesy Michael Lenihan.)

Modern entrance bridge to UCC.

Pillars of the Jury's Island railway bridge.

Today's modern arched entrance, along with the striking bridge that spans the Lee's South Channel, was completed and opened in 1929.[311] It is only a short few metres to the next bridge on the channel, Donovan's Bridge of 1902. These two bridges together form a very pleasing picture and indeed a contrast of bridge types, at close quarters.

As the channel flows onwards towards the sea, the area to the south, looking down on these gently flowing waters, is perhaps the oldest and most historic part of Cork, situated where the monastic founders of the city first settled. Here the channel divides into two, the main section turning slightly towards the north and up from which rise the pillars of a now long-gone railway bridge, across which ran steam trains from a railway station that once stood where the River Lee Hotel stands today. The smaller section of the channel flows between the grounds of St Aloysius Secondary School and the Crawford Technical School, turning back to the main South Channel of the river again just below the hotel. All that can be said about the lesser section of the channel is that it is a wilderness of undergrowth and discarded rubbish, a shame on our city and a place that should be cleaned and turned into a riverbank walkway or some other form of amenity for the city's people to enjoy.

This division of the South Channel forms an island on which stands the aforementioned hotel as well as a number of other businesses. These are only accessible by bridge and two join this island to the larger island that is the City of Cork. These are the Lancaster Quay Bridges and among the interesting features on the newest of them, constructed with the

Railway Bridge from Jury's Island in 1906. Only the pillars remain today. (Courtesy Michael Lenihan.)

Pedestrian bridge between Western Gate Building and Brookfield Medical Science Complex at UCC.

redevelopment that saw the older Jury's Hotel replaced with the modern River Lee Hotel, are effigies on the southern abutments distinctly similar to those of Neptune and the sea goddesses on St Patrick's Bridge.

Picturesque entrance to UCC.

Flood-waters swamp bridges at Lancaster Quay in December 2009.

The river now approaches the oldest part of Cork and soon passes beneath the bridge that is named after the city's patron saint, St Finbarr's Bridge of 1999. It had taken more than 1,400 years for the citizens of Cork to have a bridge named after the city's founder and patron. The next bridge we meet is one of great antiquity, Clarke's Bridge of 1776. Along the channel from here, where Beamish and Crawford brewed their ales for over two centuries, stood the old walls that enclosed the City of Cork. Yet more turnings and over another weir and there, in splendour where it has stood for three centuries is Cork's oldest bridge still in use today, the South Gate Bridge. Here one can feel the very essence of history; in front, behind, to the north and the south are memories of bygone days, of people, places and all that went to make Cork what it is today.

By contrast, Nano Nagle Bridge, the next bridge on the South Channel is a modern construct, dating from 1985. A short distance downstream from this stands another of the city's older bridges, Parliament Bridge. By now, being right in the heart of the city, the river has lost some of its peaceful qualities, but that is not to say that the bridges cannot be appreciated, especially if they are enjoyed in quieter times such as early mornings or Sunday afternoons.

Trinity Bridge of 1977 is the next spanning the river's South Channel. It is another footbridge located on a bend on the river as it begins to turn back towards its rejoining with the North Channel. With the College of Commerce on the left and the newly

The most westerly bridge on the South Channel of the Lee. Part of the Banks of the Lee Walk.

rebuilt School of Music on the right, the next bridge is the most recent Parnell Bridge, built in 1971 and alongside which is the beautiful building of the city's administration, the City Hall. Clontarf Bridge soon follows, seen by some as an ugly structure but one that has a special place in the hearts of Cork people, giving memories of a bygone age when steam trains puffed across the channel here and shipping passed through the uplifted bridge. Finally, on the South Channel is Éamon de Valera Bridge, built in 1984, after which the two channels converge again like old friends meeting after a spell apart, glad to share their travels and their stories with one another.

Eighteen bridges span the South Channel of the Lee including the two oldest bridges in use in the city, seven footbridges (one of which, Nano Nagle Bridge, is the only spanning the river to be named after a woman), and a variety to equal many other places, a fact that is continued as we journey back up the North Channel.

Journeying upstream, the first two bridges on the North Channel are twins of the final two on the South Channel; Michael Collins Bridge, built in conjunction with Éamon de Valera Bridge; and Brian Boru, in conjunction with Clontarf, is the second half of the railway pair. Upstream of these two bridges is the very heart of the shopping centre of Cork. Here stands St Patrick's Bridge, the flagship of the fleet, the iconic bridge that stands in the gaze of the Fr Mathew Statue and which joins the city's main street to the north side of the city. The river is wide here, flowing between two limestone walls and spanned next by a bridge that is young in age but old in conception in the city's family of bridges; Christy Ring Bridge, built in 1987 on the site of an ancient ford. The city's Opera House towers on one side while on the other Mulgrave Road leads to the city's north-western hills. Further upstream and on the northern bank the beautiful St Mary's church, Pope's Quay, just beyond which is one of Cork's twenty-first century bridges, Shandon Bridge, beneath the famous steeple.

Next is the North Gate or Griffith's Bridge at the site of the northern entrance to the old walled city. This bridge continues the chain of many bridges that have stood at the site through the centuries. St Vincent's is one of four footbridges that span the North Channel of the river and stands at the foot of Wyse's Hill near the entrance to the Irish Distillers plant that occupies an island in the channel. The area of this island was once known as Reilly's Marsh, and spanning that section of the North Channel that forms the island along with the main channel proper, is another of the city's oldest bridges, It was once known as Reilly's Bridge and dates to the early eighteenth century. Today it is more commonly referred to as Wyse's or the Distillery Bridge. Also crossing this lesser mill race section of the North Channel that flows to the north of the Irish Distillers island is another footbridge, one in use by those who work on the island. There had been yet another crossing point further along, but that has now disappeared beneath the car parks that have developed on the island.

In the early years of the twenty-first century the pedestrian/cycle amenity developed by the corporation begins on the North Mall and then journeys on through

Distillery Island along the bank of the river. At the western end of the island, another of the city's newer bridges, Mardyke Bridge, brings the walker across the river channel onto the Mardyke Walk from where there is access to Fitzgerald's Park or further, to the Lee fields on the Carrigrohane Straight.

Just upstream from this western end of Distillery Island is perhaps the river's most beautiful and serene section. With Fitzgerald's Park on the south bank and the gardens of the houses of Sunday's Well on the north, the hustle and bustle of the city centre is now far behind. The river's only suspension bridge, Daly's Bridge of 1927, spans here, sparkling white and beautiful in the morning or evening sunshine. The last bridge on the North Channel, just downstream from where the division beneath County Hall occurs, is Thomas Davis Bridge, though many will refer to it as Wellington Bridge.

These then are the bridges on the North Channel, ten on the channel proper and a further two on the old distillery mill-race sub-channel. These twelve, along with the eighteen on the South Channel, complete the family of thirty bridges that span the waters of the Lee within the confines of the city. Scherzer Rolling Lift Bascule bridges, limestone bridges, a suspension bridge. Cork has them all. There is nothing nicer than to walk these bridges on a quiet afternoon and contemplate such things as the stone of St Patrick's Bridge that encapsulates the glass jar with the account of the bridge's opening, the walls of Clarke's Bridge that have stood for nearly 250 years and the walls of South Gate Bridge that have stood for even longer.

At the outset of this book it was suggested that there are four fundamental characteristics in the defining of a bridge: it facilitates social and economic interactivity; is an adornment on the hinterland; is a testimony to the historical circumstances from which it emanated, and, finally, that it can be a stage upon which the rituals of the power-brokers are enacted. In telling the story of Cork's River Lee bridges, it is clear that these defining characteristics hold true.

For centuries, the bridges that stood at the North and South Gates of the walled town connected the spinal North and South Main Streets with the surrounding hinterland, encouraging urban rural interconnection. As drawbridges they served to act as part of the town's defences. The building of stone bridges early in the eighteenth century was symbolic of a new sense of security as well as a statement of expansionary intent on the part of the city authorities. The significance of their construction within a quarter century of the damage inflicted on the city walls during the siege of 1690 cannot be under-estimated.

During the eighteenth century, the city expanded eastwards onto the marshy islands that were then joined together through the infilling of the river channels that ran between them, forming the streets of the modern city. The building of the first St Patrick's Bridge in 1789 brought the building of St Patrick's Street to completion and also provided a route to the developing north-eastern residential areas to where many of the city's merchants were gravitating. The opening ceremony was a series of rituals by the Masonic Order and the city's officials which were acclaimed by the populace when they responded in kind to the cheers of the participating dignitaries.

The Masonic Order were also heavily involved in the opening of the George IV Bridge of 1820, again with all the ceremonial trappings and again the occasion was a reaffirmation of the Order's importance in society. As part of a new major route westwards, this bridge also encouraged interconnectivity between the city and the western reaches of the county and beyond.

The Anglesea/Parnell Bridge story has, as a starting point, the enabling of the economic developments in the city. When the first bridge came to be replaced in the second half of the nineteenth century, the debate surrounding what it should be called reflected the changing political dynamics of the time. At the opening ceremony, the Mayor and the city officials sought a low-key event but the crowds acclaimed Parnell and floated a balloon high above the city stating their position.

In 1902, the Lord Lieutenant and his entourage performed the opening of a bridge that had been built as part of a residential development at Donovan's Road. The procession and ceremonies associated with this event symbolised that for many in Cork, despite the emerging sense of nationalism, the trappings of royal association were still part of their lives.

The railway bridges of 1912 enabled the interconnection of the rail services of the county with those of the rest of the country. Naming the bridges Clontarf and Brian Boru recalled the great event of 1014 when the Viking invaders were defeated and banished from Ireland and choosing these names occurred precisely at a time when an emerging nationalism sought to reassert an independent Irish identity and contributed to the reawakening of a sense of a great Gaelic past in pursuit of that goal.

Two bridges of the 1980s were named in honour of the men who led Ireland in her fight for that identity but who then led opposing factions in a bitter Civil War. Divided in history but united now in a reconciling gesture that pointed the way to a new and better future, Collins and de Valera were the last bridges to be named with political overtones.

The developments of the first decade of the twenty-first century reflected the Europeanization of Irish society. The designs of the newest bridges mirrored architecture elsewhere in the country and beyond. Nevertheless, the ceremonies performed and the names chosen for the bridges reflected a sense of place and identity, a pride in the history of the city.

The future holds much for Cork and the city's bridges will play their part in that future. It is intended to redevelop the site where the Beamish and Crawford brewery once stood at the South Gate Bridge and included in the plans are two further pedestrian bridges across the Lee's South Channel. There are also plans to redevelop large sections of the industrial areas to the east of the city, where once stood giants of industry in Cork's past, Ford's, Dunlop's and Goulding's and this will necessitate more bridges being built. These, however, will be downstream of the twin channels that form the island city. Progress and onward movement for the city and people of

Cork, but let's hope it doesn't change too much of Cork and that the city's bridges will be preserved and remain, that the children of this generation and their children in turn, can discover their past by crossing Cork's greatest legacy ... that of her bridges.

NOTES

1. Raymond Gillespie and Myrtle Hall (eds), *Doing Irish Local History; Pursuit and Practice* (Institute of Irish Studies: Queen's University Belfast, 1998), Introduction.

2. Granted the downstream tunnel that was opened on Friday 21 May 1999 was named after another political figure, former Taoiseach Jack Lynch, but that was in honour of his being an iconic Cork man first and foremost.

3. See M. Phillips and A. Hamilton, 'Project History of Dublin's River Liffey Bridges', Proceedings of the Institution of Civil Engineers, *Bridge Engineering 156*, Issue BE 4 (December 2003), pp 161-79.

4. D.B. Steinman, *Famous Bridges of the World* (Dover Publications Inc; New York, 1953), p. 10.

5. David Bennett, *The Creation of Bridges* (Aurum Press Ltd: London, 1999), p. 6.

6. Peter O'Keefe and Tom Simmington, *Irish Stone Bridges, History and Heritage* (Irish Academic Press Ltd: Dubin, 1991).

7. Michael Barry, *Across Deep Waters* (Frankfort Press: Dublin, 1985)

8. M. Phillips and A. Hamilton, 'Project History of Dublin's River Liffey Bridges', Proceedings of the Institution of Civil Engineers, *Bridge Engineering 156*, Issue BE 4 (December 2003), pp 161-79; Fred Hammond, *Bridges of Offaly County: An Industrial Heritage Review* (Offaly County Council, November 1995).

9. Máire and Liam de Paor, *Early Christian Ireland* (London, 1965), p. 100.

10. *The Four Masters Annals of the Kingdom of Ireland from the Earliest Times to the Year 1616*, third edition, Vol. 1. (Dublin, 1990), p. 287.

11. Evelyn Bolster, *A History of the Diocese of Cork from the Earliest Times to the Reformation*, (Shannon, 1972), p. 38; Donnchadh Ó Corráin, 'Prehistoric and Early Christian Ireland' in Roy Foster, (ed.), *The Oxford Illustrated History of Ireland* (London, 1991), p. 14.

12. Evelyn Bolster, *Diocese of Cork*, p. 41; Revd John Ryan, *Irish Monasticism, Origins and Early Development* (Dublin and Cork, 1931), p. 316.

13. Ó Corráin, *Prehistoric and Early Christian Ireland* (London, 1991), p. 33.

14. Liam DePaor, 'The Age of the Viking Wars' in T.W. Moody and F.X. Martin (eds), *The Course of Irish History* (Cork/Dublin, 1980), p. 97; Ó Corráin, *Prehistoric and Early Christian Ireland*, p. 40.

15. DePaor, p. 102; Edmund Curtis, *A History of Ireland from the Earliest Times to 1922* (London, 2002), p. 30.

16. Henry Alan Jefferies, *Cork Historical Perspectives* (Dublin, 2004), p. 40-1.

17. John Bradley and Andrew Halpin, 'The Topographical Development of Scandinavian and

Anglo-Norman Cork' in Patrick O'Flanagan and Cornelius G. Buttimer (eds), *Cork History and Society* (Dublin, 1993), p. 15.

18. Brian Graham, 'Urbanisation in Ireland during the High Middle Ages', in Terry Barry (ed.), *A History of Settlement in Ireland* (London, 2000), p. 128.

19. William O'Sullivan, *The Economic History of Cork City from the Earliest Times to the Act of Union* (Cork, 1937), pp. 24-6.

20. Jefferies, *Cork Historical Perspectives*, p. 70.

21. Francis H. Tuckey, *Cork Remembrancer* (Cork, 1980), pp 18/27.

22. Jefferies, *Cork Historical Perspectives*, p. 124.

23. Revd C.B. Gibson, *The History of the County and City of Cork, Vol. 2* (London, 1861), p. 82.

24. Jefferies, *Cork Historical Perspectives*, p. 126.

25. Seamus Pender, *A Census of Ireland circa 1659* (Dublin, 1939), p. 191.

26. For a detailed account of the Seige of Cork see Diarmuid Ó Murchadha, 'The Siege of Cork in 1690' in *JCHAS* Vol. XCV, No. 254 (1990).

27. O'Sullivan, *Economic History*, p. 294.

28. One example, that of Dunscombe, is dealt with in *JCHAS*, Vol. X, 2nd series, 1904, pp. 128-31.

29. *Hibernian Chronicle*, 30 July 1787. The notice was signed by the Mayor Samuel Rowland Esq.

30. Tuckey, *Cork Remembrancer*, p. 122.

31. See for example *A Plan of the City of Cork in the year 1750* in Smith's *History* and Connor's map of 1774, Local Studies Department, Cork City Library.

32. Tuckey, *Cork Remembrancer*, p. 83.

33. *Ibid.*, p. 84.

34. *Ibid.*, p. 107.

35. *Ibid.*, p. 108.

36. Richard Caulfield, *The Council Book of the Corporation of Cork, from 1609-1643 and from 1690–1800* (Surrey, 1876), p.339.

37. *Ibid.*, p. 364.

38. *Ibid.* Tarras was a ground basaltic rock which was mixed with lime to form a hydraulic mortar and was imported from Holland. Source, Peter O'Keefe and Tom Simington, *Irish Stone Bridges, History and Heritage* (Dublin1991), p. 223.

39. Tuckey, *Cork Remembrancer*, p. 125.

40. O'Keefe & Simington, *Irish Stone Bridges*, p. 225. Examination of the underside of the bridge clearly indicates the two halves of the structure.

41. *Ibid.*

42. Gina Johnson, *The Laneways of Medieval Cork* (Cork, 2002), p. 51.

43. Tuckey, *Cork Remembrancer*, pp 128, 129, 160, 165, 170.

44. *Ibid.*, p. 138.

45. *Ibid.*, p. 42.

46. The Dictionary of Irish Architects website tells us that George Randall was the other person who may have been involved in the bridge's construction. www.dia.ie, cited 12 March 2009.

47. *Hibernian Chronicle*, Monday – Thursday 2-5 September 1776.

48. *Cork Mercantile Chronicle*, Monday, 10 March 1806.

49. O'Keefe and Simmington, *Irish Stone Bridges*, p. 226.

50. *Hibernian Chronicle*, 30 July 1787.

51. Tuckey, *Cork Remembrancer*, p. 170.

52. John Fitzgerald (The Bard of the Lee), 'Cork is the Eden for You, Love and Me', *Legends, Ballads and Songs of the Lee* (Cork, 1913), p. 2.

53. Many works detail the city's economic expansion at this time. See for example, William O'Sullivan, *The Economic History of Cork City from the Earliest Times to the Act of Union* (Cork, 1937); David Dickson, *Old World Colony* (Cork, 2005).

54. Antóin O'Callaghan, *'Cork's St Patrick's Street; A History'* (Collins Press: Cork, 2010).

55. *Hibernian Chronicle*, 16 May 1785.

56. T.F. McNamara, *A Portrait of Cork* (Cork, 1981), p. 77.

57. *Holden's Triennial Directory, Fourth Edition, for 1805, 1806, 1807* (London, 1807).

58. An excellent account of the events surrounding the destruction and subsequent reconstruction of the first St Patrick's Bridge can be read in the *Cork Examiner* reportage of the opening of the second St Patrick's Bridge, in the edition of Thursday 12 December 1861.

59. T.F. McNamara, *A Portrait of Cork* (Cork, 1981), p. 209.

60. *Cork Examiner*, 12 December 1861.

61. *Hibernian Chronicle*, 1 October 1789.

62. Catherine Bell, *Ritual, Perspectives and Dimensions* (Oxford University Press; Oxford, 1997), pp 128-9.

63. David Cannadine, 'Introduction: Divine Right of Kings' in *Rituals of Royalty: Power and Ceremonial in Traditional Societies,* David Cannadine and Simon Price (eds) (Cambridge University Press: Cambridge, 1987), pp 17, 19.

64. Nuala Johnson, www.sfu.ca/medialab/archive/2007/487/Resources/Readings/Johnson_mapping.pdf, (cited 8 August 2010).

65. Francis Tuckey, *Tuckey's Cork Remembrancer* (Cork, 1980), p. 217 says that the tolls of St Patrick's Bridge were sold for a period of twelve months at public auction on 7 September 1801, the sum involved being £1,400.

66. *Cork Mercantile Chronicle*, 13 August 1806.

67. *Cork Mercantile Chronicle*, 10 September 1806.

68. *Cork Mercantile Chronicle*, 13, 15, 19 August 1806.

69. O'Sullivan, *Economic History*, p. 352.

70. T.F. McNamara, *A Portrait of Cork* (Cork, 1981), p. 33.

71. *Ibid.*, p. 242.

72. *Cork Constitution*, 7 April 1849.

73. *Cork Examiner*, 4 May 1849.

74. *Cork Examiner*, 2 November 1853, gives a detailed account of the events under the headline 'Fearful Inundation in the City: Destruction of Patrick's Bridge and Loss of Life' [*sic*].

75. *Cork Examiner*, 4 November 1853.

76. *Cork Constitution*, 10 November 1853.

77. *Ibid.*

78. *Cork Examiner*, 7 December 1853.

79. *Cork Examiner*, 13 May 1859.

80. *Cork Examiner*, 13 January 1854.

81. *Cork Examiner*, 18 January 1854.

82. *Cork Examiner*, 28 January 1854.

88. *Cork Examiner*, 15 March 1854.

84. The announcement of the corporation's intention to seek replacement was posted in the *Cork Examiner* of 18 November 1853. A letter to the editor in the edition of 21 November demanded that swivel bridges be constructed to allow for shipping to once more travel upstream of the St Patrick's Bridge site. The author of the letter, one William Fitzgibbon of London, claimed that the removal of the facility to allow shipping to travel upstream of the bridge discriminated against businesses located on those quays.

85. *Cork Examiner*, 12 December 1861.

86. A general meeting of Builders and Stone-Masons took place at Mary Street on Monday 19 April 1858 at which the members made their feelings known that stone should be the material used in the bridge construction project. See *Cork Examiner*, 21 April 1858.

87. *Cork Examiner*, 5 May 1854.

88. *Cork Examiner*, 10 May 1854.

89. *Cork Examiner*, 14 July 1854.

90. A full copy of the report appears in *Cork Examiner*, 28 January 1854.

91. *Cork Examiner*, 20 April 1855.

92. *Cork Examiner*, 25 April 1855.

93. *Cork Examiner*, 4 May 1855.

94. *Cork Examiner*, 1 October 1855.

95. *Cork Examiner*, 30 November 1855.

96. *Cork Examiner*, 5 December 1855.

97. *Ibid.*

98. *Cork Examiner*, 7 January 1856.

99. *Cork Examiner*, 5 March, 9 May 1856.

100. *Cork Examiner*, 29 March 1858.

101. *Cork Examiner*, 9 April 1858.

102. *Cork Examiner*, 21 April 1858.

103. *Cork Examiner*, 17 June 1859.

104. *Cork Examiner*, 22 August 1859.

105. *Cork Examiner*, 11 November 1859.

106. *Ibid.*

107. *Ibid.* There was also placed in the cavity 'a specimen of every coin in the realm'.

108. The silver work was done by Mr Hawkesworth, Silversmith, Grand Parade.

109. *Cork Examiner*, 12 December 1861.

110. *Cork Examiner*, 23 April 1858.

111. *Cork Constitution*, 26 May 1859.

112. Much of the story of the rebuilding of St Patrick's Bridge and the footbridge built by Barnard is told humorously in a poem by John Fitzgerald, Bard of the Lee entitled 'The Bridge that Barnard Built'.

113. *Cork Examiner*, 25 May 1859.

114. *Cork Examiner*, 12 December 1861, as mentioned previously, gives a detailed account of events in the bridge story.

115. T.F. McNamara, *A Portrait of Cork* (Cork, 1981), p. 210.

116. Antóin O'Callaghan, *The Lord Mayors of Cork, 1900-2000*, (Inversnaid Publications: Cork, 2000)

117. T.F. McNamara, *A Portrait of Cork* (Cork, 1981), p. 210.

118. Fintan Lane, *In Search of Thomas Sheahan; Radical Politics in Cork 1824-1836* (Dublin, 2001), p. 15.

119. *Cork Constitution*, 6 October 1831. Note that the arithmetic is out by a figure of 10.

120. Tuckey, *Cork Remembrancer* (Cork, 1980), p. 219.

121. *Ibid.*, p.223.

122. Roger Herlihy, *A Walk through the South Parish* (Cork, 2010), p. 76.

123. See Chapter 5, for an outline of damage to bridges during the Civil War in 1922.

124. *Cork Examiner*, 5 November 1930.

125. Thomas Bartlett, *The Fall and Rise of the Irish Nation, The Catholic Question 1690-1830*, (Dublin, 1992); Maura Cronin, *Country, Class or Craft?* (Cork, 1994).

126. *Freeman's Journal*, 24 July 1821.

127. *Ibid.*

128. *Freeman's Journal*, 28 July 1821.

129. Tim Cadogan and Jeremiah Falvey, *A Biographical Dictionary of Cork* (Dublin, 2006), p. 72.

130. Angela Fahy, 'The Spatial Differentiation of Commercial and Residential Functions in Cork City 1787-1863' in *Irish Geography*, Vol. XVII (1984), pp. 18,19.

131. Angela Fahy, 'Residence, Workplace and Patterns of Change: Cork 1878-1863' in Buttel and Cullen, *Cities and Merchants, French and Irish Perspectives in Urban Development, 1800-1900* (Dublin, 1986), pp 47, 48. This City Gaol is the one that became known later as the Women's Gaol.

132. *Cork Mercantile Chronicle*, 2 January 1804. See the story of St Vincent's Bridge in a later chapter.

133. *Cork Mercantile Chronicle*, 9 March 1804.

134. Tuckey, *Cork Remembrancer* (Cork 1980), p. 126.

135. Samuel Lewis, *A Topographical Dictionary of the Parishes, Towns and Villages of Cork City and County* (Cork, 1998), p. 153. Original edition, London, 1837.

136. Tim Cadogan and Jeremiah Falvey, *A Biographical Dictionary of Cork* (Dublin, 2006), p. 77.

137. This is different from the Gaol at Sunday's Well.

138. Lewis, *A Topographical Dictionary*, p. 153.

139. Harold Bagust, *The Greater Genius? A Biography of Marc Isambard Brunel* (Surrey, 2006), p.85.

140. *Ibid.*, p.105. Marc Isambard Brunel was the father of the arguably even more famous engineer Kingdom Isambard Brunel.

141. *Ibid.*, p.85.

142. According to T.F. McNamara, *A Portrait of Cork* (Cork 1981), p.211, the portico was built with the bridge

143. David Dickson, *Old World Colony* (Cork, 2005), p. 380.

144. *Ibid.*, p.382

145. *Ibid.*

146. T.F. McNamara, *A Portrait of Cork* (Cork 1981), p.212.

147. Cadogan and Falvey, *A Biographical Dictionary of Cork* (Dublin, 2006), p. 79.

148. *Cork Constitution*, 5 June 1830.

149. Joseph Lee, *The Modernisation of Irish Society 1848-1918* (Dublin, 2008), p. 74.

150. *Cork Examiner*, 12 October 1863.

151. *Cork Examiner* 15 October 1863.

152. *Cork Examiner*, 2 December 1863.

153. *Cork Examiner*, 3 December 1863.

154. *Cork Examiner*, 5 November 1869. The committee consisted of Messrs Haynes, Johnson, C.B. Egan, Young and Morgan.

155. *Cork Examiner*, 13 November 1869.

156. *Cork Examiner*, 6 October 1874.

157. *Cork Examiner*, 9 October 1874.

158. *Ibid.*

159. Cadogan and Falvey, *Biographical Dictionary of Cork*, p.194.

160. See later in this chapter.

161. The formal designation for this Act is 1875: 38 & 39 Victoria Ch. CIX.

162. *Cork Examiner*, 4 July 1876.

163. *Cork Examiner*, 8 July 1876.

164. *Cork Examiner*, 3 July 1877.

165. Thomas Claxton Fiddler was a London engineer, born around 1841, who became Professor of Engineering at University College Dundee. In 1887 he wrote *A Practical Treatise on Bridge Construction*. http://www.dia.ie/architects/view/, cited 5 March 2009.

166. *Cork Examiner*, 11 March 1879.

167. Notice of seeking tenders appeared in the local press. See *Cork Examiner* 10 June 1879.

168. *Cork Examiner*, 7 June 1879.

169. Rooney's original tender price was £3,029 19*s* 8*d*.

170. *Cork Examiner*, 10 June 1879.

171. The other tenders came from John Steel and Sons, Cork - £2,860 8*s*; Terence O'Flynn, £3,291 8*s* 9*d*; John Wodley, Ballymoney Co. Antrim but this was unsigned and so was dismissed.

172. *Cork Examiner*, 21 June 1879.

173. *Cork Examiner*, 26 June 1879.

174. *Cork Examiner*, 9 October 1879.

175. *Irish Builder*, Vol. XXIII, 1881, p. 50.

176. *Cork Examiner*, 13 August 1879.

177. These were discussed at a number of corporation meetings at this time. See for example Friday 3 March 1882 when a deputation from the Ratepayers Association met the corporation seeking to know whether penal clauses would be invoked. They also enquired whether the corporation would insist on the original specifications being adhered to, following reports they had received that changes had been made in the designs and drawings of the turntable of the bridge. These changes had been brought to the attention of the Bridge Committee by the Harbour Board Engineer Mr Barry. A letter to the Bridge Committee from Claxton Fiddler was read to the corporationin which he said that any changes had been submitted to the committee and approved by it in August 1880. He said that there would always arise in the erection of such works circumstances and considerations which call for continual exercise of the Engineer's judgement and which necessitates in many instances departure from the lines of the drawings. He would attend the corporation in the near future to further explain his decisions.

178. Joep Leersen, *Hidden Ireland, Public Sphere* (Galway, 2002). Leersen uses the German concept of a ridge between two different eras to explain the emergence of a new concept of Irishness throughout the nineteenth and early twentieth centuries.

179. *Cork Examiner*, 29 July 1882.

180. At the 28 July meeting Ald. Nagle stated that at a meeting in his ward about a year previously, 'the subject was fully discussed and the sense of the meeting was thoroughly that it should be called Parnell Bridge ...' See *Cork Examiner*, 29 July 1882.

181. Those who voted for St Finbarr's Bridge were Messrs Geary, Gibbings, Hegarty, Harris and Smith. For Parnell Bridge were Messrs Keller and Linehan.

182. *Cork Examiner*, 15 July 1882.

183. See above.

184. For a detailed overview of this phenomenon as it developed in Dublin see Yvonne Whelan, *Reinventing Modern Dublin: Streetscape, Iconography and the Politics of Identity* (Dublin, 2003).

185. *Cork Examiner*, 2 August 1882.

186. *Cork Examiner*, 18 November 1882.

187. *Cork Examiner*, 24 November 1882.

188. Among these issues were the delay by the Stockton Forge Company in delivering iron work and changes made in the plans to the approaches to the bridge work by Claxton Fiddler, changes which Rooney maintained were unnecessary, ill considered and contributed to him running into financial difficulties and not being in a position to complete them. Furthermore, he believed, changes to the contract clauses made it easier to remove him from the project.

189. *Cork Examiner*, 19 September 1882.

190. *Cork Examiner*, 30 September 1882.

191. *Cork Examiner*, 7 May 1856.

192. *Cork Examiner*, 27 November 1857.

193. *Cork Examiner*, 1 March 1858.

194. *Cork Examiner*, 14 December 1857.

195. *Cork Examiner*, 9 August 1861.

196. *Cork Examiner*, 22 November 1861.

197. *Cork Examiner*, 8 March 1862.

198. *Cork Examiner*, 13 June 1862.

199. *Cork Examiner*, 20 June 1862.

200. *Cork Examiner*, 25 June 1862.

201. *Cork Examiner*, 17 July 1862.

202. *Cork Examiner*, 17 July 1862.

203. *Cork Examiner*, 11 August 1862.

204. *Cork Examiner*, 1 September 1862.

205. *Cork Examiner*, 22 September 1862.

206. *Cork Examiner*, 30 September 1862.

207. *Cork Examiner*, 1 October 1862.

208. *Cork Examiner*, 16 January 1863. Barry McMullen was a renowned builder who was responsible for many magnificent churches among his works. These included Thurles Cathedral, St Mary's in Cork City and the church of St Mary and St John in Ballincollig, Co. Cork, where a stained-glass window was erected in his honour.

209. *Cork Examiner*, 3 March 1863. The four local tenders were Perrott and Sons, £5,300; Steel and Co. £5,100; Henry and Charles Smith, £4,975; Cork Steam Ship Company, £4,350.

210. Newspapers in April of that year do not carry the story of the laying of the foundation stone. However, in a report of the opening of the bridge on 17 March 1864, the *Cork Examiner* states that 'the foundation stone was only laid during the month of April last'.

211. *Cork Examiner*, 16 March, 23 April 1863.

212. *Cork Examiner*, 23 September 1863.

213. A detailed description of the specifications of the bridge was reported after its opening in March 1864.

214. *Cork Examiner*, 17 March 1864.

215. *Ibid.*

216. *Cork Examiner*, 18 March 1864.

217. *Cork Examiner*, 19 March 1864.

218. *Cork Mercantile Chronicle*, 2 January, 9 March 1804.

219. *Cork Examiner*, 1 March 1858; 25 May 1859.

220. *Cork Examiner*, 9 August 1861.

221. *Cork Examiner*, 1 February 1862.

222. This church was opened in 1856 add see O'Callaghan, Sandra, *St Vincent's Church, Sunday's Well History and Heritage*, Sandra O'Callaghan Digital Art and Design (Cork, 2012).

223. *Cork Examiner*, 2 February 1875.

224. *Cork Examiner*, 5 June 1877.

225. *Cork Examiner*, 8 December 1877.

226. *Cork Examiner*, 23 February 1878.

227. *Cork Examiner*, 24 May 1878.

228. *Cork Examiner*, 27 May 1877.

229. *Cork Examiner*, 22 November 1892.

230. Yvonne Whelan, *Reinventing Modern Dublin* (Dublin, 2003), p. 99.

231. Voussoirs are blocks in which the upper edge is wider than the lower edge and in bridge building, are set flank to flank to create an arch. The central block is the keystone.

232. *Cork Examiner*, 8 May 1902.

233. *Ibid.*

234. *Ibid.*

235. *Cork Constitution* reported that the bridge was opened in the presence of an influential assembly and also that the amount contributed by the Corporation was £700.

236. *Cork Examiner*, 21 December 1899.

237. *Cork Examiner*, 13 May 1905.

238. *Cork Examiner*, 5 July 1905.

239. *Cork Examiner*, 15 May 1905.

240. The Fishguard and Rosslare Railways and Harbours Act of 1898 was mentioned in the *London Gazette* on 16 August 1898.

241. *Cork Examiner*, 14 March 1906.

242. *Cork Examiner*, 29 March 1906.

243. House of Commons debate, 1 May 1906, Vol. 156 cc488-98 as detailed in Hansard: House of Commons daily debates. http://www.publications.parliament.uk/pa/cm/cmhansrd.htm cited 28 March 2009. The reference to tramways is regarding plans to connect the city's electric tramway to join the railway system. For a detailed account of the tramway system see Walter McGrath, *Tram Tracks through Cork* (Tower Books, Cork, 1981).

244. *Ibid.*

245. *Cork Examiner*, 3 April 1906.

246. House of Commons debate, 26 June 1906, Vol. 159 cc891 - 4 as detailed in Hansard: House of Commons Daily debates. http://www.publications.parliament.uk/pa/cm/cmhansrd.htm cited 28 March 2009.

247. Walter McGrath, 'A Tale of Two Bridges' in *Cork Examiner*, 5 February 1980.

248. This company was involved in many famous bridge building projects including the Severn Bridge and the Wye Bridge. It was founded in 1877.

249. *Cork Examiner*, 12 April 1911.

250. For a more detailed description of the supporting columns and the depths etc. to which they were placed in the river beds see 'Description of the Work', *Cork Examiner*, 13 December 1911.

251. *Cork Examiner*, 30 December 1911.

252. *Cork Examiner*, 4 January 1912, 13 January 1912.

253. See *Cork Examiner*, 30 August, 1, 9 November, 22 December 1927, 10 January, 22, 23 February, 3, 8, 9 March, 12 April, 10 May 1928.

254. T.F. McNamara, *Portrait of Cork* (Cork, 1981), p. 211.

255. For a detailed history of the Cork Link Railway, including elements of the Clontarf and Brian Boru Bridge stories, see Colm Creedon, *The Cork, Bandon and South Coast Railway* in three volumes, (Cork, 1986, 1989, 1991). See also Stanley C. Jenkins, *The Cork, Blackrock and Passage Railway* and

The Cork and Muskerry Light Rail (Oxford; 1992, 1993).

256. *Cork Examiner*, 14, 15 August 1922.

257. *Freeman's Journal*, 12 August 1922.

258. *The Irish Times*, 12 August 1922.

259. *Cork Examiner*, 16 August 1922.

260. *Cork Examiner*, 19 August 1922.

261. See Chapter 3 above.

262. *Cork Examiner*, 26 April 1910.

263. *Ibid.*

264. *Cork Examiner*, 16 June 1926.

265. For a detailed account of the rebuilding of Cork's St Patrick's Street see O'Callaghan, *Cork's St Patrick's Street; A History* (Collins Press: Cork, 2010), Chapter 9.

266 Aodh Quinlivan, *Philip Monahan A Man Apart* (Institute of Public Administration: Dublin, 2006), p. 60.

267. *Ibid.* pp 59–69.

268. *Cork Examiner*, 1 November 1924.

269. *Cork Examiner*, 1 May 1925. The Carroll's Dock Bridge did not span the main North Channel of the Lee but was located just between the end of Mulgrave Road and Camden Quay. Following a report to Philip Monahan by the City Engineer on 12 May 1924, at a sitting of the Commissioner on 21 October, Monahan announced that he was seeking a loan to have the estimated £2,500 reconstruction project undertaken. The bridge was removed to facilitate that section of the Land Utilisation and Transportation project that included the building of the Christy Ring Bridge in the 1980s.

270. *Cork Examiner*, 29 July 1926.

271. *Cork Examiner*, 14 August 1926.

272. *Cork Examiner*, 3 February 1927.

273. *Cork Examiner*, 22 February 1927.

274. *Cork Examiner*, 11 April 1927.

275. In this report City Engineer Farrington is given the initials F.W. whereas previously he had been called S.W.

276. T.F. McNamara, *Portrait of Cork* (Cork, 1981), p. 214.

277. *Ibid.*

278. Dáil Eireann debates, Vol. 151, 8 June 1955. Sourced from http: historical debates. oireachtas.ie/d/151/d.0151.195506080028.html; cited 30 March 2009.

279. Dáil Eireann debates, Vol. 157, 30 May 1956. Sourced from http: historical-debates. oireachtas.ie/d/157/d.0157.195605300026.html; cited 30 March 2009.

280. *Cork Examiner*, 11 January 1968.

281. *Cork Examiner*, 11 January 1968.

282. *Cork Examiner*, 19 January 1968.

283. Dáil Eireann debates, Vol. 232, 1 February 1968. Sourced from http: historical-debates. oireachtas.ie/d0232/d.0232.196802010028.html; cited 30 March 2009.

284. *Cork Examiner*, 25 May 1971.

285. *Cork Examiner*, 15 October 1977.

286. Pat Poland, *For Whom the Bells Tolled* (The History Press Ireland: Dublin, 2010), Chapter 1.

287. Skidmore, Owings & Merrill; Martin & Voorhees Associates; E.G. Pettit & Co.; Roger Tym & Associates. *Cork Land Use and Transportation Study* (Cork, 1978), p. 89.

288. *Ibid.*, p. 206.

288. *Cork Examiner*, 20 November 1984.

290. *Cork Examiner*, 15 June 1985.

291. *Cork Examiner*, 11 July 1863.

292. *Cork Examiner*, 30 November 1863.

293. *Cork Examiner*, 4 March 1864.

294. *Cork Examiner*, 12 March 1864.

295. *Ibid.*

296. *Cork Examiner*, 29 August 1864.

297. *Constitution or Cork Advertiser*, 15 July 1870.

298. *Cork Examiner*, 14 February 1987.

299. *Ibid.*

300. For a detailed account of the siege of Cork see Diarmuid Ó Murchadha, 'The Seige of Cork, 1690', in *Journal of the Cork Historical and Archaeological Society* (1990).

301. *Evening Echo*, 7 December 1999. The *Irish Independent* described it as 'a multi-million pound new bridge' in the 7 December edition.

302. *Irish Times*, 7 December 1999.

303. J.S. Crowley, R.J.N. Devoy, D. Linehan, P.O'Flanagan, *Atlas of Cork City* (Cork, 2005), Preface, XI.

304. See for example letters in the *Cork Examiner*, 4 and 12 March 1864.

305. This architectural firm, owned by Michael McGarry and Siobhán Ní Éanaigh is based in Drogheda, Co. Louth.

306. http://www.cork2005.ie/about2005/capitalprojects.shtml says that the cost was €2.3 million with €635,000 coming from the millennium fund.

307. *Cork City Development Plan 2004* (Cork, 2003), Chapter 10, p. 170.

308. Fehilly Timoney is an Irish consulting engineering company, founded in 1990, which specialises in energy, environment, waste and civil infrastructure. Gifford Consulting Engineers is a United Kingdom company. Together they form Fehilly Timoney Gifford, for the undertaking of a range of engineering projects. In 2006, Fehilly Timoney Gifford was awarded the Public Service Excellence Award for the Blackash Park and Ride in Cork and in the same year were finalists in the President's Award for Excellence for the Mardyke Bridge project.

309. http://www.fehilytimoney.ie/docs/mardyke.html, cited 5 April 2009.

310. *Cork City Development Plan 2004*, (Cork, 2003), Chapter 10, p. 170.

311. John A. Murphy, *The College, A History of Queen's / University College Cork* (Cork, 1995), p.112, 389, 391. Details of the design of the new entrance and who was responsible for it are to be found in the *Cork Examiner*, 21 October 1929.

Bibliography

Newspapers and Periodicals

Cork Constitution

Cork Evening Echo

Cork Examiner

Cork Mercantile Chronicle

Freeman's Journal

Hibernian Chronicle

Irish Times

Journals and Official Publications

Cork City Development Plan 2004, (Cork, 2003)

Dáil Eireann debates, Vol. 151, 8 June 1955

Dáil Eireann debates, Vol. 157, 30 May 1956

Dáil Eireann debates, Vol. 232, 1 February 1968

Irish Builder, Vol. XXIII (1881)

Irish Geography, vol. XVII (1984)

Journal of the Cork Historical and Archaeological Society, (J.C.H.A.S.) Various

Phillips, M. and Hamilton, A. 'Project history of Dublin's river Liffey bridges', Proceedings of the Institution of Civil Engineers, *Bridge Engineering* 156, Issue BE4, (December 2003)

Skidmore, Owings & Merrill; Martin & Voorhees Associates; E.G. Pettit & Co.; Roger Tym & Associates, *Cork Land Use and Transportation Study* (Cork, 1978)

Publications

Bagust, Harold, *The Greater Genius? A Biography of Marc Isambard Brunel* (Ian Allan Publishing: Surrey, 2006).

Barry, Michael, *Across Deep Waters* (Frankfort Press: Dublin, 1985).

Barry, Terry (ed.), *A History of Settlement in Ireland* (Routledge: London, 2000).

Bartlett, Thomas, *The Fall and Rise of the Irish Nation, The Catholic Question 1690-1830* (Gill & MacMillan: Dublin, 1992).

Bell, Catherine, *Ritual, Perspectives and Dimensions* (Oxford University Press: Oxford, 1997).

Bolster, Evelyn, *A History of the Diocese of Cork from the Earliest Times to the Reformation* (Irish

University Press: Shannon, 1972).

Bennett, David, *The Creation of Bridges* (Aurum Press Ltd: London, 1999).

Bradley, John and Halpin, Andrew, 'The Topographical Development of Scandinavian and Anglo-Norman Cork' in Patrick O'Flanagan and Cornelius G. Buttimer (eds), *Cork History and Society* (Geography Press: Dublin, 1993).

Buttel, P. and Cullen, L. (eds) *Cities and Merchants, French and Irish Perspectives in Urban Development, 1800-1900* (Dublin, 1986).

Cadogan, Tim and Falvey, Jeremiah, *A Biographical Dictionary of Cork* (Four Courts Press: Dublin, 2006).

Cannadine, David, 'Introduction : Divine Right of Kings' in *Rituals of Royalty: Power and Ceremonial in Traditional Societies*, David Cannadine and Simon Price (eds), (Cambridge University Press: Cambridge, 1987).

Cannadine, David and Price, Simon (eds), *Rituals of Royalty: Power and Ceremonial in Traditional Societies* (Cambridge University Press: Cambridge, 1987).

Caulfield, Richard, *The Council Book of the Corporation of Cork, from 1609-1643 and from 1690–1800* (Surrey, 1876).

Creedon, Colm, *The Cork, Bandon and South Coast Railway* in three volumes (Cork, 1986, 1989, 1991).

Cronin, Maura, *Country, Class or Craft? The Politicisation of Skilled Artisans in Nineteenth Century Cork* (Cork University Press: Cork, 1994).

Crowley, J.S., Devoy, R.J.N., Linehan, D., O'Flanagan, P. (eds) *Atlas of Cork City* (Cork University Press: Cork, 2005).

Curtis, Edmund, *A History of Ireland from the Earliest Times to 1922* (Routledge: London, 2002).

De Paor, Liam, 'The Age of the Viking Wars' in Moody, T.W. and Martin, F.X. (eds), *The Course of Irish History*, (RTÉ/Mercier Press: Dublin/Cork, 1980).

De Paor, Máire and Liam, *Early Christian Ireland* (Thames and Hudson: London, 1965).

Dickson, David, *Old World Colony* (Cork University Press: Cork, 2005).

Fahy, Angela, 'The spatial differentiation of commercial and residential functions in Cork City 1787-1863' in *Irish Geography*, vol. XVII (1984).

----------------- 'Residence, workplace and patterns of change: Cork 1787-1863' in Buttel, P. and Cullen, L., *Cities and Merchants, French and Irish Perspectives in Urban Development, 1800-1900* (Dublin, 1986)

Fitzgerald, John (The Bard of the Lee), 'Cork is the Eden for You, Love and Me', *Legends, Ballads and Songs of the Lee* (Guy and Co.: Cork, 1913).

Gibson, Revd C.B. *The History of the County and City of Cork, Vol. 2* (London, 1861).

Gillespie, Raymond and Hall, Myrtle (eds), *Doing Irish Local History; Pursuit and Practice* (Institute of Irish Studies: Queen's University Belfast, 1998).

Graham, Brian, 'Urbanisation in Ireland during the High Middle Ages' in Terry Barry (ed.), *A History of Settlement in Ireland* (Routledge: London, 2000).

Hammond, Fred, *Bridges of Offaly County: An Industrial Heritage Review* (Offaly County Council, 1995).

Herlihy, Roger, *A Walk through the South Parish* (Cork, 2010).

Holden's Triennial Directory, Fourth Edition, for 1805, 1806, 1807 (London, 1807).

Jenkins, Stanley C., *The Cork, Blackrock and Passage Railway* (The Oakwood Press: Oxford, 1992).

----------------- *The Cork and Muskerry Light Rail* (The Oakwood Press: Oxford, 1993).

Jeffries, Henry Allen, *Cork Historical Perspectives* (Four Courts Press: Dublin, 2004).

Johnson, Gina, *The Laneways of Medieval Cork* (Cork City Council: Cork, 2002) .

Lane, Fintan, *In Search of Thomas Sheahan; Radical Politics in Cork 1824-1836* (Irish Academic Press: Dublin, 2001).

Lee, Joseph, *The Modernisation of Irish Society 1848-1918* (Gill & MacMillan: Dublin, 2008).

Leersen, Joep, *Hidden Ireland, Public Sphere* (Galway, 2002).

Lenihan, Michael, *Pure Cork* (Mercier Press: Cork, 2011).

Lewis, Samuel, *A Topographical Dictionary of the Parishes, Towns and Villages of Cork City and County* (Collins Press; Cork, 1998) (Original edition, London, 1837) .

McGrath, Walter, *Tram Tracks through Cork* (Tower Books: Cork, 1981).

McNamara, T.F., *Portrait of Cork* (Watermans: Cork, 1981).

Moody, T.W. and Martin, F.X., *The Course of Irish History*, (RTÉ/Mercier Press: Dublin/Cork, 1980).

Murphy, John A., *The College, A History of Queen's University College Cork* (Cork University Press: Cork, 1995).

O'Callaghan, Antóin, *Cork's St Patrick's Street; A History* (Collins Press: Cork, 2010).

----------------- *The Lord Mayors of Cork, 1900-2000* (Inversnaid Publications; Cork, 2000).

Ó Corráin, Donnchadh, 'Prehistoric and Early Christian Ireland' in Roy Foster (ed.), *The Oxford illustrated History of Ireland* (BCA: London, 1991).

O'Flanagan, Patrick and Buttimer, Cornelius G. (eds), *Cork History and Society* (Geography Press: Dublin, 1993).

O'Keefe, Peter and Simmington, Tom, *Irish Stone Bridges, History and Heritage* (Irish Academic Press Ltd: 1991).

Ó Murchadha, Diarmuid, 'The Siege of Cork in 1690' in *J.C.H.A.S.* Vol. XCV no. 254, 1990.

O'Sullivan, William, *The Economic History of Cork City from the Earliest Times to the Act of Union* (Cork University Press; Cork, 1937).

Pender, Seamus, *A Census of Ireland circa 1659* (Dublin, 1939).

Poland, Patrick, *For Whom the Bells Tolled* (The History Press Ireland: Dublin, 2010).

Quinlivan, Aodh, *Philip Monahan A Man Apart* (Institute of Public Administration: Dublin, 2006).

Ryan, Revd John, *Irish Monasticism, Origins and Early Development* (Talbot Press: Dublin and Cork, 1931).

Rynne, Colin, *The Industrial Archaeological of Cork City and its Environs*, Dúchas (The Heritage Service: Government of Ireland, 1999).

Steinman, D.B. *Famous Bridges of the World* (Dover Publications Inc.: New York, 1953).

Tuckey, Francis H., *Cork Remembrancer* (Tower Books: Cork, 1980).

The Four Masters Annals of the Kingdom of Ireland from the Earliest Times to the Year 1616, Third Edition, Vol. 1 (Dublin, 1990).

Whelan, Yvonne, *Reinventing Modern Dublin: Streetscape, Iconography and the Politics of Identity* (UCD Press: Dublin, 2003).

If you enjoyed this book, you may also be interested in…

Haunted Cork

DARREN MANN

Discover the darker side of Cork with this collection of spine-chilling tales from the archives of the Paranormal Database. Featuring stories of unexplained phenomena, apparitions, poltergeists, changelings and banshees and including accounts of mysterious vanishing islands, ghosts of shipwrecked Spanish sailors, and, of course, the story behind the legendary Blarney Stone, this book contains many spooky narratives that are guaranteed to make your blood run cold.

978 1 84588 694 3

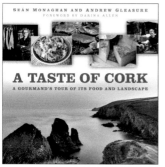

A Taste of Cork: A Gourmand's Tour of its Food and Landscape

ANDREW GLEASURES

Wild and beautiful, *A Taste of Cork* takes the reader on a voyage around a stunning county on the very edge of Europe. Cork's rich and diverse landscape is renowned, but in this book one of Ireland's top photographers, Seán Monaghan, presents it in a new light, combining the spectacular vistas with the world of the artisan gourmet food producers who are so much a part of the culture.

978 1 84588 714 8

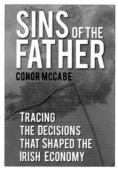

Sins of the Father

CONOR MCCABE

The questions surrounding how the Irish economy was brought to the brink have been rightly debated at length. But beyond this very legitimate exercise, there are deeper questions that need to be answered. These questions relate to why we made the decisions we did, not just in the last 10 years, but over the last 80. How did certain industries become prominent at the expense of others, banking as opposed to fisheries, international markets as opposed to indigenous industry and job creation?

978 1 84588 693 6

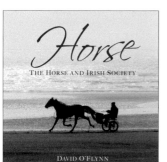

Horse: The Horse and Irish Society

DAVID O'FLYNN

Throughout the ages the horse has played a pivotal role in Irish society. It is a ubiquitous symbol in Ireland, present in many classic depictions of our culture, from loyal work horse to traditional horse fairs. While focusing on a central theme, David O'Flynn has collected an incredibly diverse number of striking photographs, ranging from elegant shots taken at the annual show jumping at the RDS to images of the milling crowd at the horse fair in Ballinasloe. Through these photographs he has created an informal and unique portrait of the Irish.

978 1 84588 706 3